International Interdisciplinary Conference "Euler Readings MRSU 2017"

Journal of Physics: Conference Series Volume 996

Moscow, Russia
22 – 24 November 2017

ISBN: 978-1-5108-6125-1
ISSN: 1742-6588

Printed from e-media with permission by:

Curran Associates, Inc.
57 Morehouse Lane
Red Hook, NY 12571

Printed by Curran Associates, Inc. (2018)

For permission requests, please contact the Institute of Physics
at the address below.

Institute of Physics
Dirac House, Temple Back
Bristol BS1 6BE UK

Phone: 44 1 17 929 7481
Fax: 44 1 17 920 0979

techtracking@iop.org

Additional copies of this publication are available from:

Curran Associates, Inc.
57 Morehouse Lane
Red Hook, NY 12571 USA
Phone: 845-758-0400
Fax: 845-758-2633
Email: curran@proceedings.com
Web: www.proceedings.com

Table of contents

Volume 996

International Interdisciplinary Conference "Euler Readings MRSU 2017"

22–24 November 2017, Moscow Region State University (MRSU), Russian Federation

Accepted papers received: 19 March 2018
Published online: 4 April 2018

Preface

International Interdisciplinary Conference "Euler Readings MRSU 2017"

Peer review statement

Papers

Introductory Paper

Development of Euler's ideas at the Moscow State Regional University

P I Vysikaylo and V V Belyaev.....1

Mathematics and Mathematical Physics

Geometric situation of points of division of regions of direct and return currency in channels with the presence of heating zone limited on the longitudinal coordinate

A Gerasimov, A Kirpichnikov and F Sabirova.....7

Mathematical modeling of the MHD stability dependence on the interpole distance in the multianode aluminium electrolyser

R N Kuzmin, N P Savenkova, A V Shobukhov and A V Kalmykov.....13

Mathematical modeling of the toroidal vortex structures interaction

S A Skladchikov, N P Savenkova and R N Kuzmin.....18

The stability issues in problems of mathematical modeling

A. Yu. Mokin, N. P. Savenkova and N. S. Udovichenko.....23

Analytical properties of nonequilibrium threshold in shock waves

M M Kuznetsov, Y D Kuleshova, Yu G Reshetnikova and L V Smotrova.....29

The H-theorem and equation of state for kinetic model of imperfect gas

A. M. Bishaev, V. A. Rikov and M. V. Abgaryan.....32

On interaction of P-waves with one-dimensional photonic crystal consisting of weak conducting matter and transparent dielectric layers

A A Yushkanov and N V Zverev.....37

On behavior peculiarity of electron plasma

N M Gordeeva and A A Yushkanov.....41

The estimation of the internal dimensions of the channel for transport of atoms, molecules from the variation principles and the conditions of its local symmetry

N A Krylova and E V Kalashnikov.....45

Physics and Material Science

Sources and components of ball lightning theory

A I Nikitin, V L Bychkov, T F Nikitina, A M Velichko and V I Abakumov.....50

On modeling of "plasmoid" created by electric discharge

V L Bychkov, S V Anpilov, N P Savenkova, V Stelmashuk and P Hoffer.....70

On electro-hydrodynamic effects over liquids under influence of corona discharge

V L Bychkov, V I Abakumov, A R Bikmukhametova, V A Chernikov and D A Safronenkov.....77

On the physics of electron transfer (drift) in the substance: about the reason of "abnormal" fast transfer of electrons in the plasma of tokamak and at known Bohm's diffusion

I A Boriev.....86

Solar wind as a consequence of Coulomb mirror effect in astrophysics

P I Vysikaylo and M A Korotkova.....92

Model of superconductivity formation on ideal crystal lattice defect–twin or twin boundary (MSC-TB)

V A Chizhov, F S Zaitsev and V L Bychkov.....101

Existence of dark matter with observed properties of cosmic microwave background radiation substantiates three conservation laws of classical physics and all principles of quantum mechanics as creates the value of Planck's constant

I A Boriev.....106

Theory of liquid crystal orientation under action of light wave field and aligning surfaces

A K Dadivanyan, D N Chausov, V V Belyaev, N N Barabanova, O V Chausova and Yu D Kuleshova.....113

Interaction of dyes CD–1 and SD–1 with the surface of oligodimethysiloxane

D N Chausov.....117

Variational Approach in the Theory of Liquid-Crystal State

E V Gevorkyan.....122

Practical Application and Economics

Controllable light beam shaper in LED lamp

K D Nessemon, I V Popov, V V Belyaev, A A Belyaev and V K Velichko.....125

Eulerian Circles (Venn Diagrams) as model for modern economy education on the basis of Russian professional standards

M A Sharonov, O V Sharonova and V P Sharonova.....130

International Interdisciplinary Conference "Euler Readings MRSU 2017"
Moscow, Russia, 22-24 November 2017

Philipp I Vysikaylo and Victor V Belyaev

Moscow Region State University, Radio street 10A, Moscow, 105005, Russia

E-mail: filvys@yandex.ru, vv.belyaev@mgou.ru

Preface

This volume of the IOP Conference Series: Materials Science and Engineering contains papers from the International Interdisciplinary Conference "Euler Readings MRSU 2017" which took place from the 22nd to the 24th November 2017 at the Moscow Region State University, Moscow, Russia.

The primary aim of this Conference was to celebrate 250th anniversary of a discovery made in Russia by the famous Swiss scientist Leonard Euler in 1767. The Conference also provides a forum where scientists and students can meet and discuss different current interdisciplinary problems in the following areas:

• Studies of structures and fields in the Space;
• Turbulence and structural turbulence in potential wells. Transport processes in inhomogeneous media and in media with potential wells;
• Structures in continuous media and their properties. Cumulative dissipative structures (linear and beaded lightnings, cathode spots, stratum, cords, plasmoids, etc.) in the plasma and their properties. Vortex structures (cyclones, tornadoes, electric cords behind meteoroids, etc.);
• Shock and detonation waves and their cumulative and dissipative properties and transformation;
• Ambipolar drift and diffusion transport processes and inhomogeneous structures in plasma;
• Methods of generalized mathematical description of analogs in various phenomena (including gravitational, electromagnetic, nuclear, acoustic and quantum phenomena);
• Doped crystals (excitons, exciton molecules, superlattices) and their properties;
• Application of Euler's ideas in the development of social sciences about social structures and their interaction.

A review of plenary and invited talks is presented in a paper by P Vysikailo and V Belyaev "Development of Euler's ideas at the Moscow State Regional University".

This proceedings volume contains the full texts of 22 papers on related topics. All the papers have been subjected to peer review.

February 2018

Acknowledgements

The International Interdisciplinary Conference "Euler Readings - MRSU 2017" has been supported by the Russian Foundation of Basic Researches (RFBR), grant No.17-07-20558-g.

Peer review statement

All papers published in this volume of *Journal of Physics: Conference Series* have been peer reviewed through processes administered by the proceedings Editors. Reviews were conducted by expert referees to the professional and scientific standards expected of a proceedings journal published by IOP Publishing.

Development of Euler's ideas at the Moscow State Regional University

P I Vysikaylo and V V Belyaev

Moscow Region State University, Radio street 10A, Moscow, 105005, Russia

E-mail: filvys@yandex.ru

Abstract. In honor of the 250th anniversary of Euler's discovery of three libration points in Russia in 1767 in the area of two rotating gravitational attractors in 2017 an International Interdisciplinary Conference "Euler Readings MRSU 2017" was held in Moscow Region State University (MRSU). The Conference demonstrated that the Euler's ideas continue to remain relevant at the present time. This paper summarizes the main achievements on the basis of Leonard Euler's ideas presented at the Conference.

1. Introduction. Leonard Euler discovery of the libration points

In 2017, international scientific community celebrates 250 anniversary of one of the most important discoveries made in Russia by the famous Swiss scientist Leonard Euler in 1767. The discovery of three linear libration points by Euler (the nonlocal interaction of various potentials - two gravitational and centrifugal) is still relevant and is used in many sciences about the Space, in the theory of gas-discharge phenomena, nanotechnologies, and other theoretical and technical sciences. This is due to the analogy of the long-range interaction of Coulomb, gravitational, centrifugal and even social potentials, such as language that comprises history, habits, customs and even the spirit of peoples acting synergistically (together) and thereby forming a joint civilization or opposing each other, which results in the destruction of the states that vacillate between them.

The fact that a libration point (oscillations of the third body between the attractors) should be between two gravitationally interacting spatially separated bodies (attractors) - L_1 (Figure 1), where the forces of attraction of the third body to these two gravitational attractors (equalizers) are equal to each other was supposed by Newton, who discovered the law of universal gravitation. Euler was the first to show that the libration point L_1, with the centrifugal potential taken into account, multiplies by three points - L_1, L_2 and L_3 (L_1 is reflected behind these bodies rotating relative to each other and results in the formation of two more linear libration points L_2 and L_3 behind them). Thus Euler transformed for the first time the Newton point to the Euler line (Figure 1). Therefore he took into account the nonlocality (the highest derivative with respect to the coordinates) - in the momentum transfer equation (the term in Euler's equation - $(V\nabla)V$) as well as showed to us methods how to solve and generalize many problems, including for describing social phenomena between two or more global social attractors.

Five years later, Lagrange emerged into the plane of rotation of the gravitating system of two bodies (the Sun and Jupiter) and the third small body and opened the triangular libration points L_4 and L_5 in front and behind Jupiter. In contrast to the Euler points these two points, proved to be absolutely Lyapunov stable with respect to weak perturbations. Then in 1904, astronomers discovered in these

IOP Conf. Series: Journal of Physics: Conf. Series **996** (2018) 012001 doi:10.1088/1742-6596/996/1/012001

libration points a system of asteroids ahead of and behind Jupiter, moving in unison along the orbit of Jupiter. After that all the points of libration and cumulation, including those discovered by Euler, were illegally called the Lagrange's points.

The escape into the 3D space from the plane of rotation of double stellar systems was carried out by MRSU employees. Earlier, some of the specific tasks for calculating centrifugal potentials have been solved by Academician E I Zababakhin in analyzing the destruction of unlimited cumulation of cumulative jets or converging rotating liquid metal shells [1] and Academician L A Artsimovich in describing the reflection of charged particles in the mirror cells [2]. All these 3D tasks were summarized in the works of the MRSU employees. This resulted in the discovery of 3D structural turbulence at the libration points and quasi-Cooper bicyclones. Such cyclones enhance the rotation of each other, although rotating in opposite directions in a single system [3] like electrons in the Cooper pair. Neither Kolmogorov, nor Obukhov, nor other researchers of turbulence could guess on the structural turbulence. They did not take into account the centrifugal potentials that describe the energy-mass-impulse flows convergent or cumulative to the center. They believed that viscosity erases turbulent vortices and rotation. Further in 2009, the MRSU employee discovered not only the libration points between two positively charged structures, but also lines and even the libration planes for electrons in the region of positively charged structures, which allowed explaining the causes of the Faraday dark space, strata and a number of phenomena during the ingoing and destruction of meteoroids into the electronegative atmosphere of the Earth. However, do not forget that it all started with the work by Euler, in which he first took into account the influence of the centrifugal potential on the structure of energy-mass impulsive flows of small bodies in the region of two rotating gravitational attractors. Therefore the 250th anniversary of this event was celebrated in MRSU on November 22-24, 2017 by the International Interdisciplinary Conference "Euler Readings - MRSU 2017" [4]. Anyone who is interested in the problems of natural sciences, technology, as well as social phenomena and music could participate in the conference.

2. Main Sections of the International Interdisciplinary Conference "Euler Readings MRSU 2017"

On the first day of the conference, many topical interdisciplinary problems were discussed:

- Research of structures and fields in the Space and application of this knowledge in practice;
- Turbulence and structural turbulence in potential wells. Transport processes in inhomogeneous media and in media with potential wells;
- Structures in continuous media and their properties. Cumulative-dissipative structures in plasma (linear and pulsating lightning, cathode spots, strates, electric cords, electric tails behind meteoroids, plasmoids, etc.), solid bodies (excitons, exciton molecules and superlattices), liquids (gutters, etc.), gases (tornadoes, cyclones, etc.) and their properties. Vortex structures (cyclones, bi-cyclones, tornadoes, tornadoes, electric tails behind meteoroids, etc.);
- Shock and detonation waves and their cumulative and dissipative properties and transformation;
- Ambipolar drift and diffusion transport processes and inhomogeneous structures in plasma;
- Methods of generalized mathematical description of analogs in various phenomena (including gravitational, electromagnetic, nuclear, acoustic and quantum phenomena);
- Alloyed crystals (excitons, exciton molecules, superlattices) and their properties;
- Development of the Euler's ideas for describing phenomena and structures in nanochemistry;
- Application of the Euler's ideas in the development of social sciences about social structures and their interaction.

Many MRSU staff members from different MRSU faculties and departments as well as many specialists from numerous Russian and international institutions took an active part in the discussion of these subjects.

Numerical and analytical solutions of problems and modeling of the processes accompanying similar phenomena in such different environments involve the use of generalized mathematical transposition methods (GMTM) or generalized transposition methods (GTM) of knowledge and mathematical models from areas well studied in the field of knowledge, currently insufficiently studied. This implies the further development of the foundations of cognitive sciences (interdisciplinary scientific direction, which forms the theory of artificial intelligence), allowing obtaining new knowledge in the fields of natural sciences on the basis of the analogues learned. In this respect, as shown by the Interdisciplinary International Conference "Euler Readings - MRSU 2017", the Euler's discoveries are still relevant and are used as a scientific foundation in many sciences about the Space, fullerenes, etc., can be used to explain phenomena in the electron-hole plasma, waves of spatial recharge of traps, and other theoretical and technical sciences. Relying on the Euler's work and methods of transferring mathematical models, one can acquire new knowledge in completely different sciences and apply them in technologies from femto- to macrocosm.

The talks were presented in a few sections. The program included many leading scientists of Russia: Academician Son E.E., Corresponding Member of the Russian Academy of Sciences D V Bisikalo, Academician S T Surzhikov, Academician A S Bugaev et al.

3. Review of the Conference Talks

In the introductory lecture of the conference "Euler Readings - MRSU 2017" Prof. P I Vysikaylo paid special attention to the problem of studying not only the gravitating systems investigated by Euler and Lagrange, but also systems with violation of neutrality. As was shown for the first time by MRSU employees besides of the libration points, discovered by Euler and Lagrange, there are libration points between the structures with the same charge sign. Properties of such structures have already been described. It is shown that in these points between positively charged structures there is the cumulation of the electric potentials of electrons, their energy and fluxes. As a result, the Faraday dark spaces are formed in the region of the libration point between the positively charged cathode spot and the positive plasma column. Consideration of a weak violation of the neutrality of structures and cumulative processes due to a violation of neutrality makes it possible to explain the stratification of discharges in gases, the Pekarik effect (the opposite direction of the group and phase velocities of perturbation propagation in a plasma with current), the reverse motion of a cathode spot in a transverse magnetic field, the constant solar wind, the electric wind, the formation and Coulomb levitation (ascent) of cylindrical columns of water or alcohol above their surface.

The investigation of the solar wind is conducted by graduate and post graduate MRSU students under supervision of MRSU Professor D.Sc. Vysikaylo P.I. A report "On electrohydrodynamic action of the corona discharge on a liquid" was presented by professor of the Moscow State University Bychkov V.L. The structurization and properties of droplets of liquids is studied under the guidance of Honorary Professor of Moscow State University and Honored Scientist of the Russian Federation Kuzmin R.N. The points of libration between positively charged crystals of phosphors and the role of physical doping of phosphor materials by nanostructures from allotropic forms of carbon were reported by Ms. Kovaliev A.I. from St.-Petersburg Technology Institute (graduate student of Professor Sychev M.M.).

As shown by the MRSU staff, the number of the libration points between atoms determines their valence in molecules and crystals. Besides, the libration points as well as libration lines and even libration surfaces for electrons arise in charged plasma structures [4]. The Debye' radius is not a panacea for all physics. And the application of the approach, where the Debye radius is the main characteristic size of all problems, is very limited. Myths and false conjectures should be verified by experiments, not by learned mantras.

The generalization of the above discoveries and studies allowed the MRSU staff to formulate a method of the generalized mathematical transposition, consisting in the adequate transfer of the mathematical models and methods of scientific description of phenomena from areas well studied and verified by numerous experiments in the field of natural sciences that have so far not been studied well enough with wrong conclusions refuted by experiments. To-date, the staff of MRSU have obtained solutions of the system of Euler and Poisson equations in perturbation theory, which made it possible to study in the plasma three types of ambipolar diffusions (Schottky, Poisson and Euler) and compare their relative contribution under varying conditions. Thus, inertia and violation of neutrality results in new types of diffusion of a quasineutral (but not absolutely neutral) plasma. On the basis of experiments with fullerenes, all hypotheses and provisions on the partial restriction of the de Broglie hypothesis in quantum mechanics are attributed to erroneous conclusions.

Professor of the Peoples' Friendship University of Russia (RUDN University) N V Samsonenko spoke on current problems of quantum mechanics. He was a scientific trainee in the de Broglie group in 1980-1981. So he was informed firsthand by the founder of quantum mechanics and his disciples. In his report "The Choice of Dynamic Variables and the Interpretation of Quantum Mechanics" jointly with co-authors Ndahayo Fidel (Department of Physics of the Rwanda University, Butare, Rwanda), Usman Manga Adam (Department of Physics, Abdu Moumouni University, Niamey, Niger), he noted that due to the specific features of physical laws in the microregions of our space-time and unusual mathematical apparatus, quantum mechanics occupies a special place among the physical sciences about the nature. To date, it is considered the most proven and most successful theory in the history of science, but there is still no consensus in understanding its deep meaning. Despite the fact that with its help we are able to calculate the values of many physical quantities with high accuracy a saying by R. Feynman is valid: "I think I can safely say that nobody understands quantum mechanic" [Feynman, Richard (1965). *The Character of Physical Law*. Modern Library]. N V Samsonenko has finished his report: "It's funny that all the experimenters with whom the author met in Russia and abroad spontaneously stand on de Broglie-Bohm positions, despite the official teaching in all universities of the world of the Copenhagen interpretation of quantum mechanics."

Professor P I Vysikaylo has made a single remark to the fundamental talk of N V Samsonenko. In his interpretation of the first axiom of quantum mechanics, there is a general mistake made by all the classics of quantum mechanics that there exists a wave function that completely describes (in the quantum sense) a physical system and its square is the probability density $\omega(r) = |\psi|^2$. In fact, in the framework of any interpretation of quantum mechanics, it is necessary to take into account the geometric normalization coefficient $\omega(r) = 2k\pi r^k |\psi|^2$ for spherical and cylindrical symmetric quantum resonators in the probability density $\omega(r)$. Here k is equal to 2 for spherical symmetry and $k = 1$ for the cylindrical symmetry of the quantum resonators. Such consideration rehabilitates the solutions of the Helmholtz equation, ejected by the requirement that Dirac's psi function is to be limited (he ejected the so-called $\cos(kr)/r$ waves with its own energy spectrum – $E_{n-1/2}$ from classical quantum mechanics). Such waves are named also as the fundamental tone. They are observed in conventional acoustics. They describe such phenomena as sonoluminescence and cavitation). The calculation of cos-waves ($E_{n-1/2}$) in hollow quantum resonators (for example, in fullerenes and standing hollow excitons of large radius) made it possible to explain the experimentally observed Raman spectra of doped crystals, establish the foundations of cumulative quantum mechanics, increase significantly the possible spectra ($E_{n-1/2}$ and E_n) of standing large-radius excitons and at least to eliminate the contradiction of the de Broglie hypothesis with Schrödinger's quantum mechanics that was truncated by Dirac's erroneous requirement about the boundedness of the psi-function everywhere. Unlimited cumulation of the waves was taken into account in acoustics, however, it was not considered in quantum mechanics owing to Dirac's requirement.

After the creation of cumulative quantum mechanics, unlimited cumulation of the psi function was possible in hollow quantum cavities. Only the inclusion of the fundamentals of cumulative quantum mechanics allows us to conclude that the de Broglie hypothesis "quantum particles behave like waves" completely corresponds to all the wave phenomena discovered in conventional acoustics. Thus, the main achievements of the quantum mechanics for describing the behavior of the quantum particles in the quantum resonators are reduced to the de Broglie hypothesis that the quantum particles behave like waves and all the mathematical models and solutions obtained for the waves are carried completely by the description of the behavior of the quantum particles.

MRSU Professor, Doctor of Sciences in Physics and Mathematics (Department of General Physics of the MRSU) E V Gevorkyan detailed on the history of the creation of Euler's calculus of variations. It was firstly published in an Euler's work in 1744 and was described in detail in the book "Elements of the Calculus of Variations" in 1766. The Euler's calculus of variations became the foundation of not only mathematics but also theoretical physics. The paper discussed the application of Euler's variational methods in modern condensed-state physics and, in particular, in the theory of the liquid-crystal state.

Professor M Y Ivanov, Chief Researcher of the P I Baranov Central Institute of Aviation Motors (CIAM) told about the integrals of the Euler equations and solitons in the hydrodynamic plasma model.

Professor A V Gerasimov in a joint report with A P Kirpichnikov, F R Sabirova (Kazan National Research Technological University) spoke about the geometrical location of the points of separation of the regions of direct and return currents in the channels in the presence of heating zones restricted in their longitudinal coordinate. Such surfaces can be attributed as the surfaces of the Euler's libration, as the Co-Chairman of the Conference P I Vysikaylo noted.

Of particular interest was the report of Professor V V Cherniy on the nature of the origin of the rings of Saturn: possible superconductivity of ice particles of rings. Here, new mechanisms of ring formation were proposed and verified.

Professor of the Institute of Space Researches (ISR) of RAS N S Erokhin presented a study of wave propagation in inhomogeneous media. The research is in co-authorship with Academician of the Russian Academy of Sciences, Chairman of the Scientific Council of the Russian Academy of Sciences on Nonlinear Dynamics V E Zakharov and ISR RAS staff N N Zolnikova, L A Mikhailovskaya). N S Erokhin talked about the reflectionless propagation of the electromagnetic waves through the plasma with small-scale structures of large amplitudes with varying relative permittivity.

A PhD and Associate Professor V N Ilyin presented a talk on environmental problems in Russia and the Moscow region. Once again he convincingly showed that the basis of cumulative processes in social phenomena are the processes of timely purification of sewage and separate utilization of garbage. His team has original developments and sufficient experience to solve these problems effectively.

The application of the method of generalized transposition allowed the conference to analyze the behavior of the United States in relation to the Russian Federation, the European Union (Germany), and China. Here the Euler's ideas about points and areas of libration (oscillations, throwing, etc.) between two social attractors have not been applied yet. Indeed, if one looks at the big Eurasia, taking into account China, India, Indonesia, Japan, then the calm synergetic development of this region will undoubtedly result in reduction of the role for the United States and its satellites. Here, the United States beat out weak links (Ukraine, Russia, Near East) with small expenses ($ 5 billion) and flood Europe with refugees from the destroyed countries. So areas of the libration are purposefully created in Ukraine, as well as problems in Russia, Germany.

A separate important event was the performance of the great-great-granddaughter Euler in the 9th generation - Hecker Nadezhda Feodorovna. She told about the descendants of Euler from his daughters in the book about the descendants of Euler and more in detail focused on the works of his father and uncle professors and famous scientists of Russia.

Undoubted adornment of the interdisciplinary conference was the speeches of the staff and students of the Yu A Shaporin Music School and the A Shnitke Moscow State Institute of Music. Their speeches occurred between scientific reports. A number of musical performances were listened to at the conference.

4. Conclusion

Despite the old anniversary - 250 years, the Euler's ideas do not become obsolete they continue to develop, bringing new results. In this case, the theory itself is practically not modified. The article is concluded with the words by Lagrange: "If you love math, then read the work of Euler", they are relevant at the moment. The full record of the conference "Euler Readings - MRSU 2017" can be viewed in youtube [4].

IOP Conf. Series: Journal of Physics: Conf. Series **996** (2018) 012001 doi:10.1088/1742-6596/996/1/012001

5. Acknowledgement

The International Interdisciplinary Conference "Euler Readings - MRSU 2017" has been supported by the Russian Foundation of Basic Researches (RFBR), grant No.17-07-20558-g.
The paper is prepared under partial support of the RFBR (grants Nos. 18-07-00897 and 17-07-00469).

References

[1] Zababakhin E I and Zababakhin I E *Phenomena of unlimited cumulation. M.*: Nauka, 1988 (in Russian) - 171 p .
[2] Artsimovich L A, *Atomic energy.* 1958. **5**. 5. P.501.
[3] Vysikaylo P I, *Engineering physics.* 2013. 3. P.3.
[4] https://youtube/5XN_hLihSac

IOP Conf. Series: Journal of Physics: Conf. Series **996** (2018) 012002 doi:10.1088/1742-6596/996/1/012002

Geometric situation of points of division of regions of direct and return currency in channels with the presence of heating zone limited on the longitudinal coordinate

A Gerasimov, A Kirpichnikov, F Sabirova

Kazan National Research Technological University 420015, Karl Marx str., 68 Kazan, Russia

gerasimov@kstu.ru

Abstract. The analysis of energy balance equation for viscous laminar flow of fluid or gas in the cylindrical channel in the area (zone) of warm up bounded along the longitudinal coordinate is made. It was found that at laminar flow of fluid or gas in a round pipe, in each warm up area bounded along the longitudinal coordinate there are the areas of direct and reverse flows separated by a plane that is a locus of points where temperature is maximal for each fixed value of radial coordinate r.

1. Introduction

In a series of papers [1 — 3] devoted to the characteristics of heat transfer and gas dynamics of a high-frequency induction plasma, in the framework of classical thermal physics a number of results concerning the gas-dynamic structure of a high-frequency induction (HFI) discharge at atmospheric pressure was obtained. In [1] the concept of fixed point of HFI discharge is introduced, where all three components of velocity of plasma gas vanish; it is also shown that it corresponds to a point on the axis of the plasmoid of HFI discharge where the value of its axial temperature is maximal at each fixed value of radial coordinate. In [3] it was found that inside the plasmoid of HFI discharge at atmospheric pressure, the areas of direct and reverse flows are separated by a surface of revolution, which is the locus of points at which the temperature of the discharge is maximum at each fixed value of the radial coordinate r. Thus, the fixed point of the high-frequency inductive discharge can be considered as a particular case of a more general physical phenomenon — the surface with zero axial velocity separating direct and reverse flows in the gas-discharge chamber of HFI plasma torch. The presence of a fixed point, drag and aft vortices in the discharge chamber of HFI plasma torch is confirmed both by the numerical results and the corresponding experimental data [4 — 10].
In [3] it was also suggested that the found phenomenon should be peculiar to a wider range of objects with internal heat source bounded along the longitudinal coordinate. In this work, this phenomenon is considered as peculiar to a rather wide class of flows in cylindrical channels in the presence in these channels of one or several zones of heating (internal or external) bounded along the longitudinal coordinate.

2. Method of solution

Let us consider the equation describing the balance of the energy of viscous laminar flow of liquid or gas inside the cylindrical channel in the area (zone) of heating bounded along the longitudinal coordinate. Ina general form it is written as follows

$$\frac{1}{r}\frac{\partial}{\partial r}\left(\lambda\, r\frac{\partial T}{\partial r}\right)+\frac{\partial}{\partial z}\left(\lambda\frac{\partial T}{\partial z}\right)+Q(r,z)=\rho c_p\left(v_r\frac{\partial T}{\partial r}+v_z\frac{\partial T}{\partial z}\right). \tag{1}$$

Here T is temperature, λ is the coefficient of heat conductivity, ρ is the density, c_p is the specific heat capacity, $Q(r,\ z)$ is the density of heat source energy put in the flow, v_r and v_z, are the radial and longitudinal components of the velocity field inside the channel, respectively. It is apparent that the component Q(r, z) differs from zero only in case if heat is put in the entire channel volume, for example by RF or microwave heating, otherwise when the pipe heating is performed by heat supply to its external surface Q(r, z) = 0. From the balance equation (1), the components considering viscous heating and compressibility of medium are excluded since their value is deliberately small compared with the values provided by external heat source.

From equation (1) it may be inferred

$$v_z(r,z)=\frac{\frac{1}{r}\frac{\partial}{\partial r}\left(\lambda\, r\frac{\partial T}{\partial r}\right)+\frac{\partial\lambda}{\partial T}\left(\frac{\partial T}{\partial z}\right)^2+\lambda\frac{\partial^2 T}{\partial z^2}+Q(r,z)-\rho\, c_p v_r\frac{\partial T}{\partial r}}{\rho\, c_p\frac{\partial T}{\partial z}} \tag{2}$$

Now by convention consider the family of cylindrical surfaces coaxial to the channel ш series filling its entire internal space and complying with certain values of radial coordinate *r* of the channel. Due to the continuity of this coordinate there will be an infinite multitude of such surfaces, and the one that agrees with the value r = 0 degenerates to the channel axis, and the one that agrees with the value r = R coincides with its wall. At each of these additional cylindrical surfaces, we distinguish a circumference that agrees with the maximal temperature of this surface bounded along the axial coordinate inside each of these heating zones, that is always possible since all heating zones are bounded in their value. It is obvious that due to the continuity of this family of surfaces the distinguished circumferences in turn form some cylindrically symmetrical surface of revolution Ω_0, which axis coincides with the axis of cylindrical channel and that may be described by the equation

$$\left.\frac{\partial T}{\partial z}\right|_{\Omega_0}=0. \tag{3}$$

Then in accordance with the methodology [3], we determine the place of the points for which condition (3) is satisfied, that is consider the surface Q~ in the channel at which points the temperature value of radial coordinate r is maximal. At all points of the surface (3), the denominator of formula (2) vanishes, and this means that the numerator of this expression shall vanish as well since the value of longitudinal velocity v, as a physical value everywhere should be finite. At the points of the surface Ω_0, due to (2), it is obvious that

$$\frac{1}{r}\frac{\partial}{\partial r}\left(\lambda r\frac{\partial T}{\partial r}\right)+\frac{\partial\lambda}{\partial T}\left(\frac{\partial T}{\partial z}\right)^2+\lambda\frac{\partial^2 T}{\partial z^2}+Q(r,z)-\rho\, c_p v_r\frac{\partial T}{\partial r}=0$$

from where the reverse component of velocity is

$$\left.v_r(r,z)\right|_{\Omega_0}=\frac{\frac{1}{r}\frac{\partial}{\partial r}\left(\lambda\, r\frac{\partial T}{\partial r}\right)+\lambda\frac{\partial^2 T}{\partial z^2}+Q(r,z)}{\rho\, c_p\frac{\partial T}{\partial r}} \tag{4}$$

Now come back to equation (2) and analyze the behavior of all its components at the transition through the surface Ω_0. It is clear that at the points of surface Ω_0, the denominator changes its sign at the passage through Ω_0, since

$$\frac{\partial T}{\partial z} > 0 \text{ at } z < z(\Omega_0) \text{ and } \frac{\partial T}{\partial z} < 0 \text{ at } z > z(\Omega_0).$$

Neither of the numerator components, however, changes its sign at the passage through the surface Ω_0: the components $\frac{1}{r}\frac{\partial}{\partial r}\left(\lambda r \frac{\partial T}{\partial r}\right)$, $Q(r,z)$, $\rho c_p v_r \frac{\partial T}{\partial r}$ — — since they depend only on temperature and its radial (rather than axial) derivatives, and the change of transverse components of velocity because of formula (4) is small; the component $\lambda \frac{\partial^2 T}{\partial z^2}$

Is negative both to the right and to the left from this surface; and the sign of the component $\frac{\partial \lambda}{\partial T}\left(\frac{\partial T}{\partial z}\right)^2$ apparently determined by the sign of the derivative $\frac{\partial \lambda}{\partial T}$ and also does not depend on the side of the surface Ω_0 on which it is considered.

3. Results and conclusion

As we can see, zone of the components of the numerator of the formula for $v_z(r,z)$ changes its sign when passing through the surface Ω_0, so the numerator keeps the sign i. e.,- where in the vicinity of the surface, that is tends to zero on the right and left from Ω_0 a with one and the same sign. This in turn means that the longitudinal velocity v,(r,z) at all points be- longing to Ω_0, changes the sign, i.e., vanishes at the points of the surface:

$$v_z(r,z)\big|_{\Omega_0} = 0$$

Physical meaning of this expression is apparently that the surface Ω_0 is the surface separating the opposite liquid or gas flows inside the heating area (zone) bounded along the longitudinal coordinate z.

Further it may be shown that the surface Ω_0, since it already exists may be only a surface. Consider the section of the meridional of such a surface AC (Fig. 1).

IOP Conf. Series: Journal of Physics: Conf. Series **996** (2018) 012002 doi:10.1088/1742-6596/996/1/012002

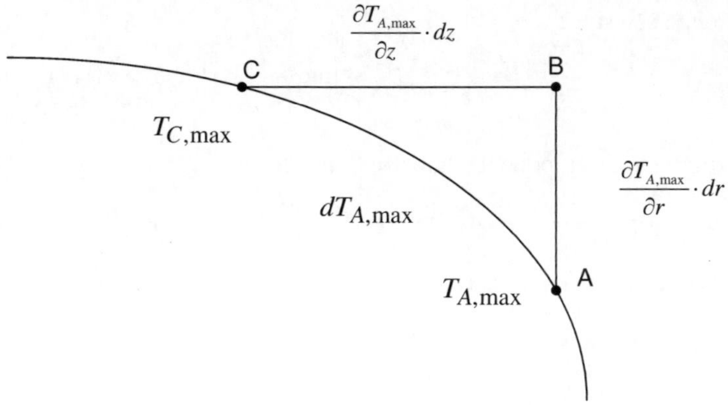

Figure 1. Meridional surface section

Anyway the temperature increment from the point A to the point C along the AC axis will be equal to

$$dT_{A,\max} = \frac{\partial T_{A,\max}}{\partial z} dz + \frac{\partial T_{A,\max}}{\partial r} dr. \tag{5}$$

But at the point A, temperature reaches its maximum along the straight line r = r_A, and, therefore, at this point

$$\frac{\partial T_{A,\max}}{\partial z} = 0.$$

Substituting the latter in equation (5), we obtain $dT_{A,\max} = \frac{\partial T_{A,\max}}{\partial r} dr$

Thus, temperature increment at the points B and C is the same, hence $T_B = T_C$. On the other hand, under the condition of maximum temperature at the point C along the line $r = r_c$, strict inequality $T_B < T_C$ should be satisfied. Failure of this condition is only possible if the points B and C coincide, that is, if the generatrix AC is the straight line, or equivalently, if the plane surface Ω_0 is the plane z = z_A. The obtained result means that the heat supplied to the heating zone (generated in this zone) propagates along the z axis on both sides relative to a plane Ω_0 mainly by convective heat transfer, and only in the periphery of the zone, the mechanism of conductive heat transfer will come into effect.

The main result obtained in the paper can be summarized as follows. In laminar flow of a fluid or gas inside a circular tube in each heating zone bounded by the longitudinal coordinate, there are areas of direct and reverse flows (Fig. 2), which are divided by the plane, which is the locus of points at which the ambient temperature $T = T$(r, z) is maximal for each fixed value of the radial coordinate r in the heating zone.

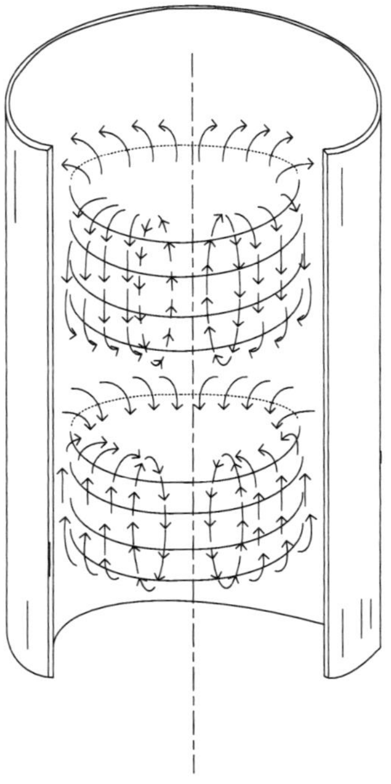

Figure 2. Kinematic flow in the heating zone

Thus, within each heating zone only a fraction of fluid and gas entering into a round pipe may be freely pumped through it (mostly along the walls). The greater part of liquid or gas is involved in the reverse recirculation motion, and at the same time, the resistance of the pipe increases dramatically. The so-called vortex locks, impeding the uniform flow of liquid or gas through a pipe are formed. This effect may be observed, of course, only at respective balance between the maximum temperature in the heating zone and the flow rate of liquid or gas supplied into the cylindrical channel. In the case, when the flow rate increases to a critical level, the vortex locks must be obviously pushed out of the heating zone. In this case, the flow has a progressive vortex-free character, and inside the round pipe, the heating zones bounded along the longitudinal coordinate are not formed any more. Gas or liquid supplied in the pipe will be evenly (and insignificantly) heated at the motion along its entire length downstream the area (zone) of heating which, thus, spreads to the entire length of the tube below its initial position.

References

[1] A.V. Gerasimov and A.P. Kirpichnikov**,** Static point of high-frequency induction discharge, High Temperature, 1999, **37**, 3, P. 504

[2] A.V. Gerasimov and A.P. Kirpichnikov, Influence of axial temperature profile on axial velocity profile in high- frequency induction discharge, Applied Physics, 2003, **39**, 3, P. 671

[3] A.V. Gerasimov and A.P. Kirpichnikov, Interface between direct and return flows in high-frequency inductive discharge, Doklady Physics, 2003, **48**, 3, P.116

[4] **V**.Kh. Goikhman and V.S. Kuzmina, On the calculation of velocity fields of gas flow in induction HF dis- charge, in: Proc. XXVII Herzen Readings, Physical Electronics, Pt. 2, Electronics of hight-temperature plasma, LSPI, Leningrad, 1974, P.66

[5] A.V. Donskoi, S.V. Dresvin, and K. El-Mikati, Gas-dynamic parameters of high-frequency induction plasmatron, in: Proc. 6th All-Rus. Conference on Low-Temperature Plasma Generators, Ilim, Frunze, 1974, P. 218

[6] S.V. Dresvin and K Kl-Mikati, Measuring and calculation of gas-dynamic parameters of induction high-frequency discharge, High Temperature, 1977, **15**, 6, P. 1158

[7] S.V. Dresvin and V.I. Borisenkov, Investigation of vortex flows in discharge chamber of HF-plasmatorch, in: VIII All-Rus. Conference on Low-Temperature Plasma, Pt. 3, Novosibirsk, 1980, P.111

[8] Philipp I. Vysikaylo Cumulative Point—L1 Between Two Positively Charged Plasma Structures (3-D Strata) IEEE Transactions on plasma science 2004 pp.1-5

[9] P. I. Vysikaylo, "Points, lines and surfaces of Vysikailo–Euler libration (cumulation) in nonuniform structures in plasma with current," in *Proc. 37th Int. Conf. Phys. Plasma CTNF*, Zvenigorod, Russia, 2010, p. 311.

[10] P. I. Vysikaylo, "Detailed elaboration and general model of the electron treatment of surfaces of charged plasmoids (from atomic nuclei to white dwarves, neutron stars, and galactic cores): Self-condensation (self-constriction) and classification of charged plasma structures—Plasmoids part 1. General analysis of the convective cumulativedissipative processes caused by the violation of neutrality: Metastable charged plasmoids and plasma lenses," *Surf. Eng. Appl. Electrochem.*, 2012. **48**, 1, p.11.

IOP Conf. Series: Journal of Physics: Conf. Series **996** (2018) 012003 doi:10.1088/1742-6596/996/1/012003

Mathematical modeling of the MHD stability dependence on the interpole distance in the multianode aluminium electrolyser

R N Kuzmin, N P Savenkova, A V Shobukhov, A V Kalmykov

Lomonosov Moscow State University, Leninskie Gory, Moscow, 119991, Russia

E-mail: mknandrew@mail.ru, shobukhov@mail.ru, alex_2391@mail.ru

Abstract. The paper deals with investigation of the MHD-stability dependence on the depth of the anode immersion in the process of aluminium electrolysis. The proposed 3D three-phase mathematical model is based on the Navier-Stokes and Maxwell equation systems. This model makes it possible to simulate the distributions of the main physical fields both in horizontal and vertical planes. The suggested approach also allows to study the dynamics of the border between aluminium and electrolyte and the shape of the back oxidation zone.

1. Introduction

An aluminium electrolyser is a complex nonlinear object. Its main volume is a rectangular bath with rounded corners (see Fig.1). At the advanced stage of the process the bath contains a layer of liquid aluminium at the bottom and a layer of melted cryolite with special additives above it. Graphite electrodes are immersed into the cryolite layer from above to a relatively small depth. In the course of the process a gas layer appears under the anodes as a result of the electrochemical reaction.

The graphite electrodes are the anodes, while the liquid aluminium and partly the bottom of the bath are the cathodes. An electric potential difference is applied to the electrodes. Meanwhile, the upper liquid layer (the electrolyte) contains positive and negative ions that are the mobile charge carriers. These ions move up and down to the corresponding electrodes and are discharged on them. As a result of these reactions, the primary aluminium and the gaseous mixture appear. The most important reactions here are the formation of various aluminium salts near the anode and the reduction of these salts to metal near the cathode, actually — at the aluminium-electrolyte border [1].

One of the most important parameters of the industrial electrolyser cell is the anode immersion depth. The greater is the depth, the smaller is the electrical resistance of the bath; thus the electrical energy consumption of the cell decreases. But at the same time the probability of the MHD instability increases, and it may cause losing the whole amount of the primary metal.

2. Description of the Problem

The proposed model is based on the so-called multiliquid approach [2,3]. It is supposed that in the heterogeneous mixture each phase occupies only a fraction of the elementary volume. In this regard we introduce the variables α_m , m=1,2,3, that characterize the volume fraction of the m-th phase in the mixture.

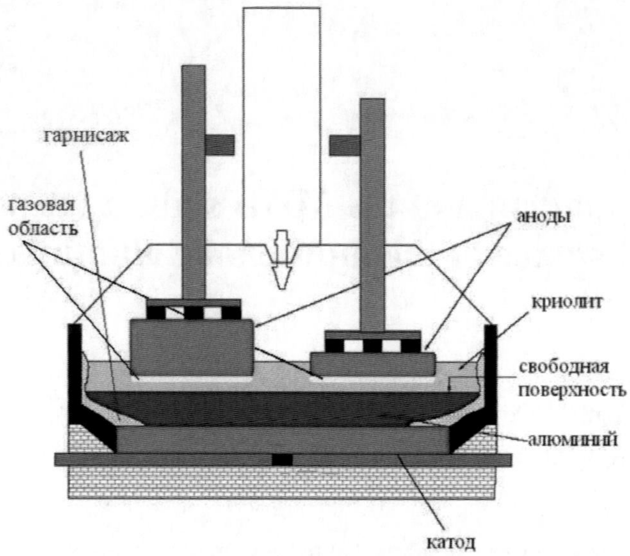

Figure 1. The cross-section scheme of the multi-anode electrolyser.

Let use denote the macroscopic velocities of the mixture components as $\mathbf{v}_m = (u_m, v_m, w_m)$ the pressure as p, the magnetic field as $\mathbf{H}_m = (H_x, H_y, H_z)$, the Lorentz force as $\mathbf{F}_m = (F_x, F_y, F_z)$, the electric current density as $\mathbf{J} = (j_x, j_y, j_z)$ and the mixture temperature as T. Then the mass variation of the m-th phase is described by the following equation:

$$\frac{\partial \alpha_m \rho_m}{\partial t} + div(\alpha_m \rho_m \mathbf{v}_m) = \mathbf{J}_m, \tag{1}$$

where ρ_m - the density of the m-th phase - is constant, and $\mathbf{J}_m$ is the source (or sink) of the m-th phase due to the chemical reactions.

The transfer of momentum for each phase is described as:

$$\frac{\partial \alpha_m \rho_m \mathbf{v}_m}{\partial t} + (\mathbf{v}_m, \nabla)(\alpha_m \rho_m \mathbf{v}_m) = -\alpha_m \nabla p + \mu_m \Delta \mathbf{v}_m + \alpha_m \rho_m \mathbf{g} + \mathbf{P}_{mn} + \alpha_m \mathbf{F}_m, \tag{2}$$

where μ_m is the dynamic viscosity of the media, $\mathbf{F}_m$ is the volume density of the Lorentz force:

$$\mathbf{F}_m = \mu_0 \left[rot\, \mathbf{H} \times \mathbf{H} \right] \tag{3}$$

and $\mathbf{P}_{mn}$ is the volume interphase viscous friction force between the n-th and the m-th mixture components.

The equation for pressure may be obtained by summation of the momentum equations for the three phases and by taking into account the balance equations:

$$\nabla \cdot \left(\sum_{m=1,2,3} \left(\frac{\alpha_m}{\rho_m} \right) \nabla p \right) = \nabla \cdot \mathbf{L}, \tag{4}$$

together with the boundary conditions:

$$\frac{\partial p}{\partial n} = (\mathbf{L},\mathbf{n}) \Big/ \sum_{m=1,2,3} \left(\frac{\alpha_m}{\rho_m} \right), \tag{5}$$

$$\mathbf{L} = \sum_{m=1,2,3} \left((\mathbf{v}_m, \nabla)(\alpha_m \mathbf{v}_m) + \frac{\alpha_m \mu_m}{\rho_m} \Delta \mathbf{v}_m + \alpha_m \mathbf{g} + \right.$$
$$\left. + \frac{1}{\rho_m} \sum_{m \neq n} \alpha_m \alpha_n (\alpha_m \mu_m + \alpha_n \mu_n)(\mathbf{v}_m - \mathbf{v}_n) + \frac{\alpha_m}{\rho_m} \mathbf{F}_m \right), \tag{6}$$

where **n** is the vector normal to the boundary.

The equation for the electric field potential is derived from the electric field divergency equation. The mixture is supposed to be electrically neutral, and thus the equation for the potential looks as follows:

$$\nabla \cdot \left(\sum_{m=1,2,3} (\alpha_m \sigma_m) \nabla \varphi \right) = \nabla \cdot \left(\mu_0 [\mathbf{v}, \mathbf{H}] \sum_{m=1,2,3} (\alpha_m \sigma_m) \right) \tag{7}$$

where **v** is the average mixture velocity, and σ_m is the conductivity of the m-th phase. The anode electric potential is set equal to 4 volts. In the nonconductive part of the bath surface the normal derivative of the electric potential equals zero. In whole the boundary conditions look as follows:

$$\varphi\big|_{анод} = 4, \; \varphi\big|_{катод} = 0, \; \left.\frac{\partial \varphi}{\partial n}\right|_{\Gamma} = 0. \tag{8}$$

We get the electric current density from the relation:

$$\mathbf{J} = \sum_{m=1,2,3} (\alpha_m \sigma_m) \left(\mu_0 [\mathbf{v}, \mathbf{H}] - \nabla \varphi \right) \tag{9}$$

which is due to the media flow, which causes induction of the additional electric field. We neglate the bias current, because the variation of the electric potential with time is small enough.

The magnetic field is the solenoidal field with the vector potential A. We find this potential using the Coulomb calibration:

$$\nabla \mathbf{A} = 0. \tag{10}$$

Using the quazistationary approximation, we get the following Neumann problem for the vector potential:

$$\begin{cases} \Delta \mathbf{A} = -\mathbf{J} \\ \left.\dfrac{\partial \mathbf{A}}{\partial n}\right|_{\Gamma} = 0 \end{cases} \tag{11}$$

Having obtained the vector potential, we can easily find the magnetic field itself.

The problem described above was solved numerically. The finite difference equations were splitted according to physical processes on the staggered meshes [2,4].

3. Results of Computational Experiments

We've carried out a series of computations that simulated the industrial process of aluminium electrolysis with anode immersion depths varying from 10 to 15 to 20 centimeter. Fig.2-4 present the dynamics of the interpole distance with respect to time.

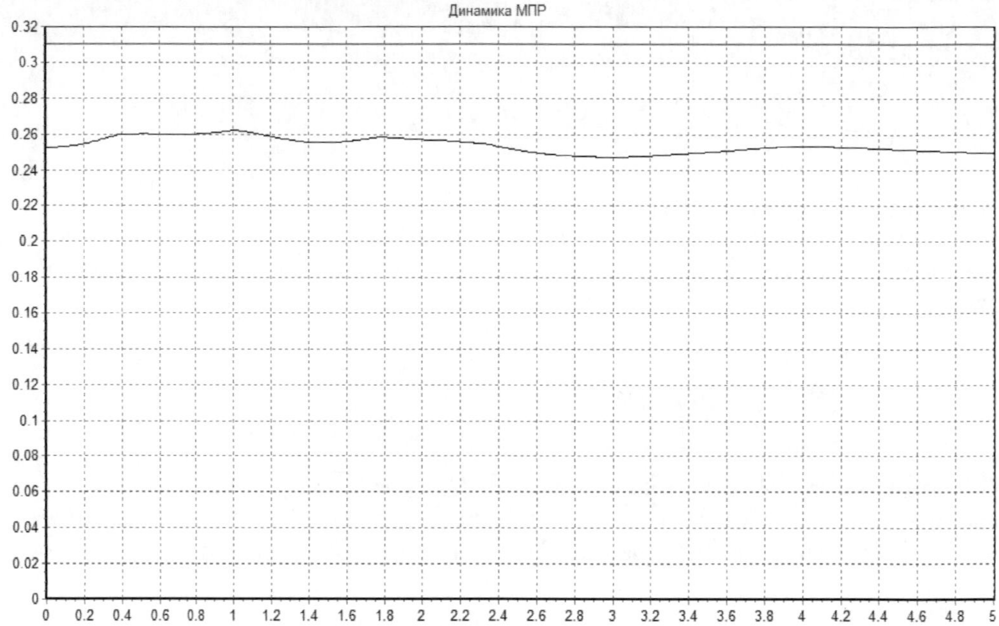

Figure 2. Dynamics of the interpole distance, anode immersion depth equals 10 cm.

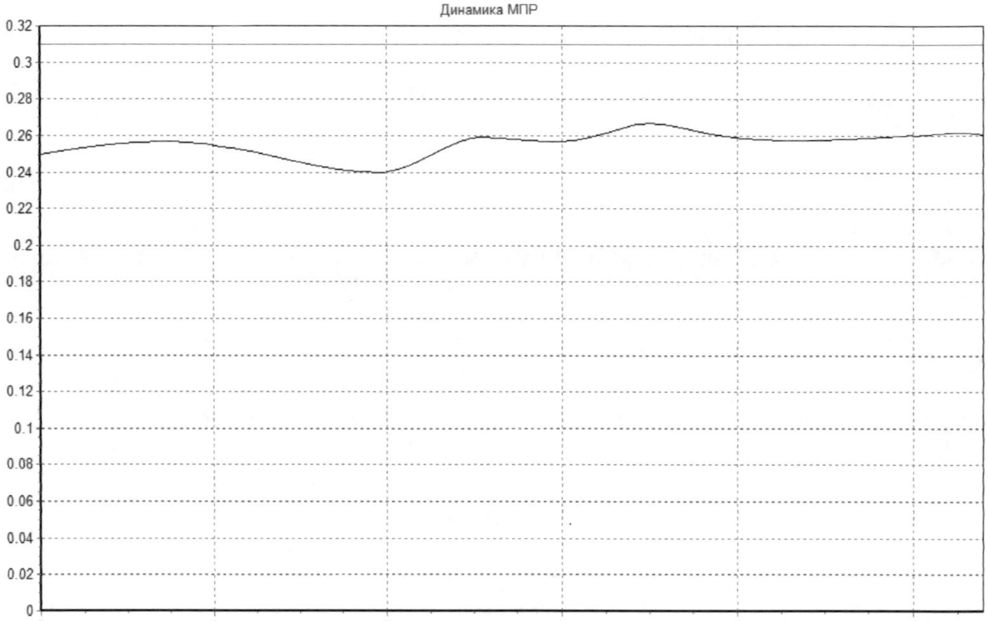

Figure 3. Dynamics of the interpole distance, anode immersion depth equals 15 cm.

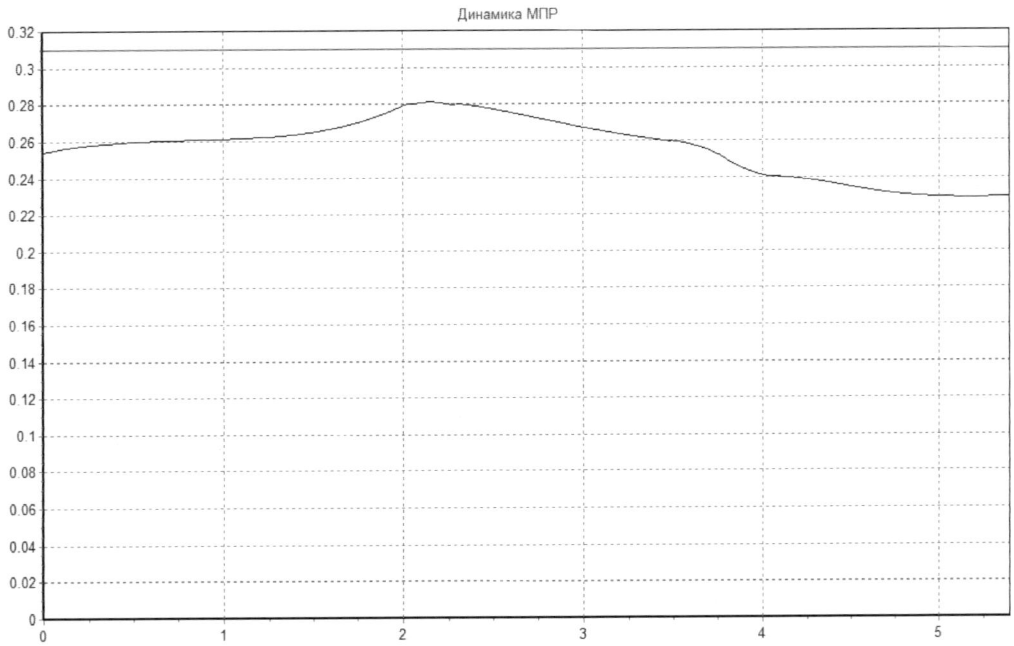

Figure 4. Dynamics of the interpole distance, anode immersion depth equals 20 cm.

Conclusion

The paper contains the results of mathematical modeling of an industrial aluminium electrolyser. We demonstrate that the amplitude of the liquid aluminium surface oscillations increases together with the depth of the anode immersion. It may cause the development of the MHD instability, but in all three cases shown here the electrolysis process is stable.

References

[1] Belolipetskii V M, Piskazhova T V 2013 Mathematical modeling of aluminium electrolysis. Solving the problems of the technology control Siberian Federal University.
[2] Savenkova N P, Anpilov S V, Kuzmin R N, Provorova O G, Piskazhova T V *A Two-phase 3D Model of Aluminium Electrolysis Cell* Theses of the Third International Congress on Non-Ferrous Metals. Kransoyarsk, 2011. 282-286.
[3] Nigmatulin R I 1978 Mechanics of Heterogeneous Media M.: Nauka
[4] Patankar S 1980 Numerical Heat Transfer and Fluid Flow Taylor & Francis

Mathematical modeling of the toroidal vortex structures interaction

Skladchikov S A [1], Savenkova N P [1], Kuzmin R N [2]

[1]Faculty of Computational Mathematics and Cybernetics, Lomonosov Moscow State University, Moscow, Russia
[2]Faculty of Physics, Lomonosov Moscow State University, Moscow, Russia

E-mail: skladtchikov@mail.ru

Abstract. The paper presents the investigation of the toroidal vortex structures (TVS) interaction in various media. The proposed mathematical model makes it possible to study in details the stages of interaction of the toroidal vortices and allows to conclude what processes occur in the real collision of vortex structures, and to predict the result of this collision.

1. Introduction

The paper presents the research of toroidal vortex structures (TVS) interaction in various media. We introduce the 3D hydrodynamic mathematical model and demonstrate the results of numerical experiments that allow us to make the conclusions listed below.

The toroidal vortex formation process has three main stages: 1) the beginning; 2) the formation of the toroidal spiral jet in emerging vortex; 3) the establishing of the azimuthal rotation velocity after termination of the inlet. The laws of motion of the toroidal vortex and the surrounding substance, not engaged in the vortex motion, become different when the solid rotational core appears inside the vortex. From this fact we derive one of the main conclusions of our research: the toroidal vortex exists as an independent hydrodynamic structure with its own laws of motion after the rotational solid core has emerged. Such vortex is stable due to the low frontal resistance coefficient of the vortex in its translatory motion through the surrounding medium. This coefficient may be dozens of times smaller than the resistance coefficient of the water drop, which is the most streamline object. The stable toroidal vortex has its own properties similar to the properties of a solid body. If we place the origin in the center of the moving vortex, the velocity field of circumfluent medium will look like the flow around the solid obstacle.

2. Mathematical model

For simulating the process of the TVS formation we use the continuity equation, the Navier - Stokes equations of motion, the heat conductivity equation and the Mendeleev-Clapeyron equation. Our mathematical model in cylindrical coordinates looks as follows:

$$\frac{\partial \rho}{\partial t}+\frac{1}{r}\frac{\partial r V_r \rho}{\partial r}+\frac{1}{r}\frac{\partial V_\varphi \rho}{\partial \varphi}=0,$$

$$\frac{\partial \rho \mathbf{V}}{\partial t}+(\rho \mathbf{V}\nabla)\mathbf{V}=-\nabla P+\mu\Delta\rho\mathbf{V}-\frac{v_0^2\rho}{r}\theta\mathbf{e}_r-2\rho[\mathbf{\Omega}\mathbf{V}],$$

$$\frac{\partial \rho\theta}{\partial t} + (\rho \mathbf{V}\nabla)\theta = \chi\Delta(\rho\theta),$$

$$V = \begin{bmatrix} V_\varphi \\ V_r \end{bmatrix}, \quad \Omega = \frac{V_\varphi}{r}, \quad \theta = \frac{(\mathrm{T}\text{-}\mathrm{T}_0)}{T_0},$$

$$\mathrm{P} = \frac{\rho}{M} RT.$$

Here V_r and V_φ are the radial and angular flow velocity components; P, T, ρ, M, μ and χ are respectfully the pressure, temperature, density, molar mass, dynamic viscosity and coefficient of thermal diffusivity; R is the universal gas constant; er is the unit vector. This model has been proposed and discussed in details in [1-6].

3. Numerical results

The TVS and the necessary conditions for its formation are described in [1-5]. Here we discuss the results of computer simulation of several TVS interactions in the gaseous and liquid media.

Let us consider the collision of two toroidal vortices in the air when both structures are formed by the same vortex generator and move in the same direction. The second vortex has higher velocity than the first one; that's why it finally overtakes the latter.

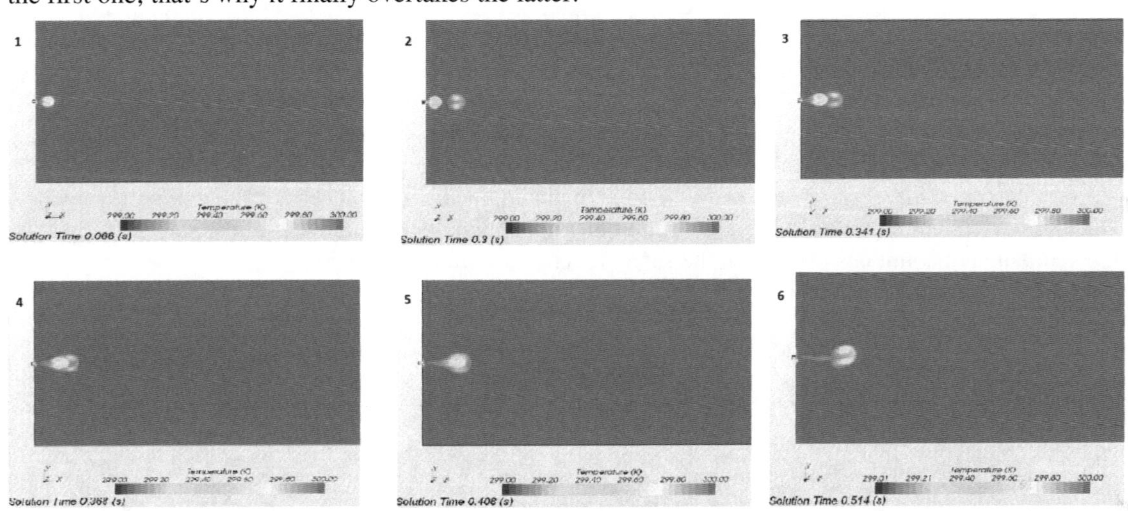

Figure 1. Collision stages for two TVS moving in the same direction

In Fig. 1 we present the process of formation and interaction of two toroidal vortices in the air. For convenience we demonstrate the evolution of the temperature distribution in time. The ambient temperature for this computation is 300 K, while the initial temperature of the emerging vortices is 299 K. The initial velocities of the first and of the second vortices are 3 m/s and 5 m/s respectively. As one can see from Fig. 1, the second vortex overtakes the first one, passes through it and then both vortices merge into a single large TVS.

Now let us consider the frontal collision of two toroidal vortices in the water. Two water structures approach each other along the horizontal axis and finally meet. The distance between the vortex generators is 1 m, and the vortices are equal in all aspects except for the initial velocity. The ambient temperature for this computation is 300 K, while the initial temperature of the emerging vortices is 299 K.

IOP Conf. Series: Journal of Physics: Conf. Series **996** (2018) 012004 doi:10.1088/1742-6596/996/1/012004

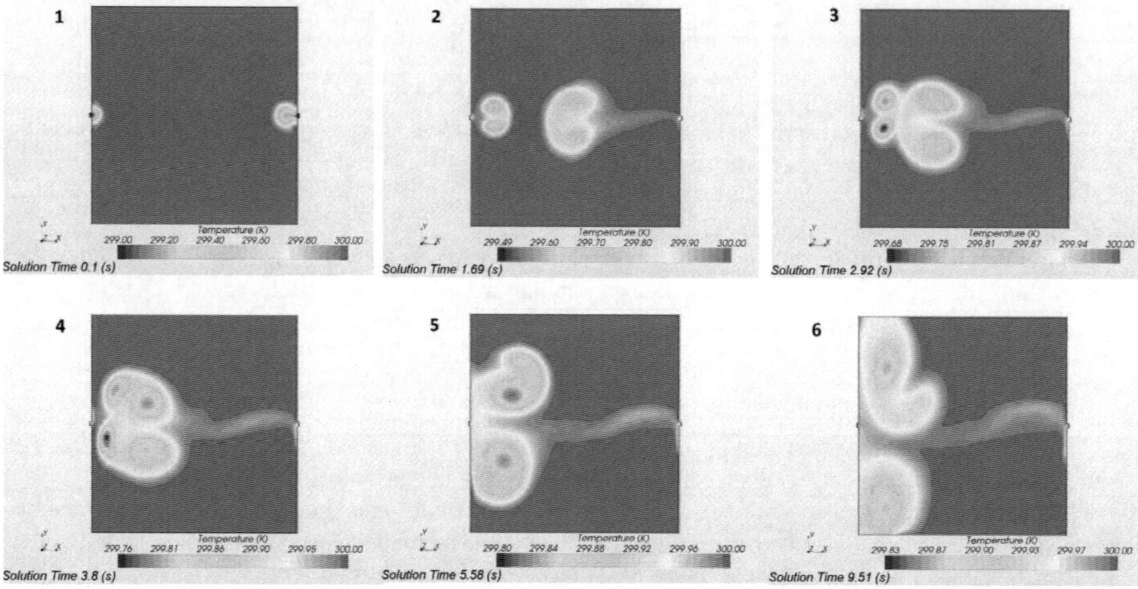

Figure 2. Collision stages for the frontal interaction of two TVS with initial velocities 1 and 3 m/s.

In Fig. 2 we present the process of formation and interaction of two toroidal vortices approaching each other. The initial velocities of the left and of the right vortices are 1 m/s and 3 m/s respectively. As one can see from Fig. 2, during the collision the right TVS, which has higher initial velocity, engages the left one into its structure. Together they make a new, more powerful and stable TVS. This unified structure continues moving in the same direction as the faster TVS.

Let us see what happens if the initial velocities are equal.

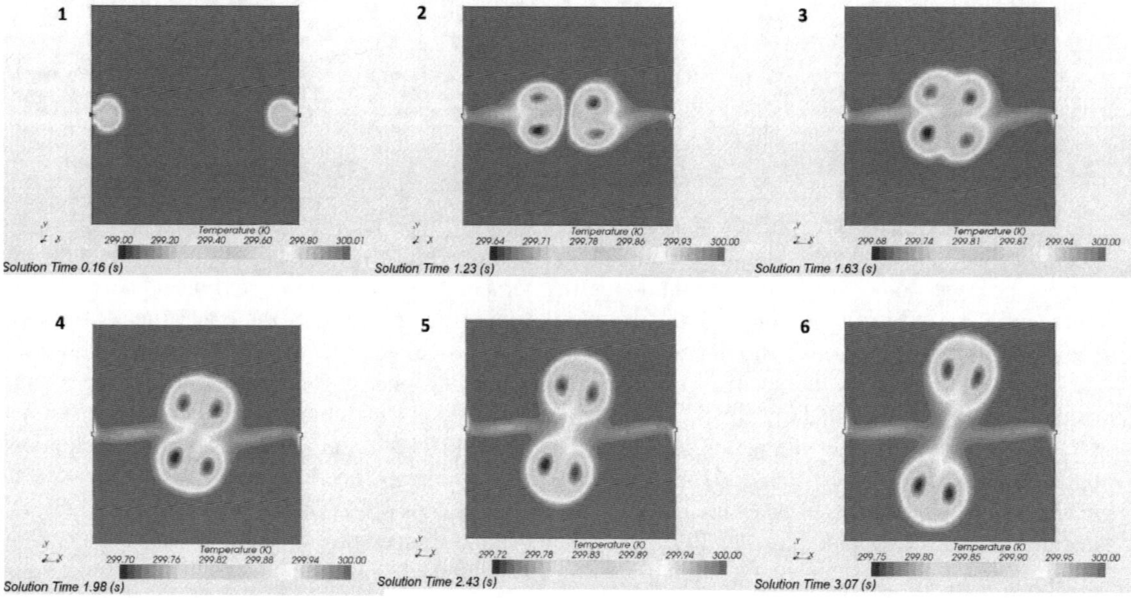

Figure 3. Collision stages for the frontal interaction of two TVS. Both initial velocities are 3 m/s.

As it is shown in Fig. 3, both vortices are equally strong and stable. That's why after the collision both TVS disappear; two new structures, moving perpendicularly to the original direction, are formed instead. This is the most interesting case, and for this reason we shall study carefully not only the temperature, but also the velocity field of both structures.

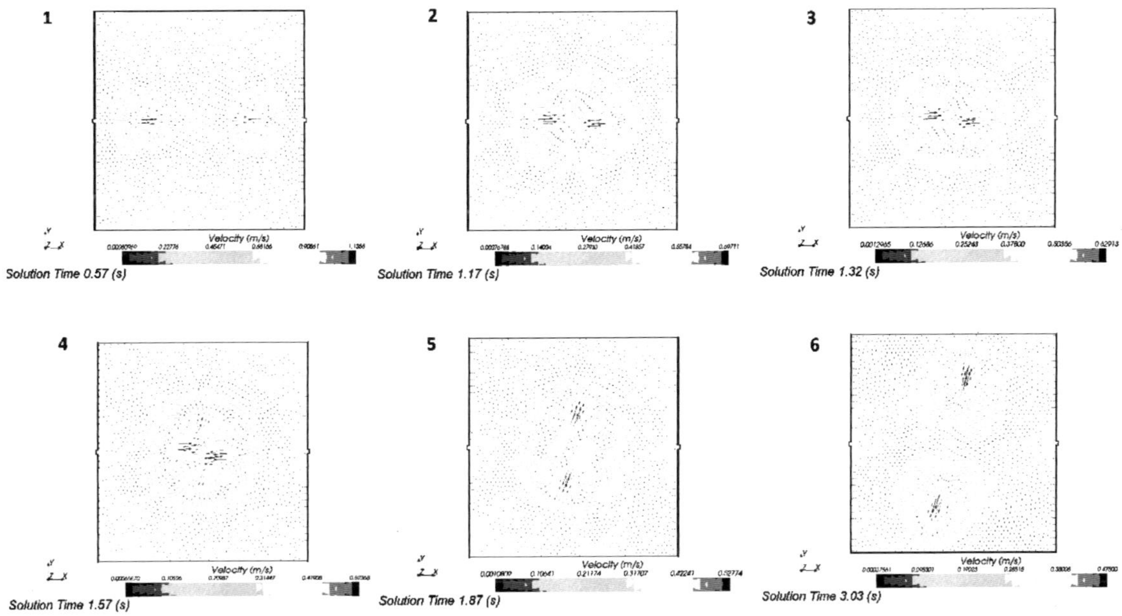

Figure 4. Velocity field of frontal collision of two TVS. Both initial velocities are 3 m/s.

In Fig. 4 we see that the maximal velocity at the moments 1 and 2 is observed in the cores of both vortices. At the moments 3 and 4 the maximal velocity changes its direction and becomes perpendicular to the initial one. It means that the vortex transformation has begun, and the new cores emerge. At the moments 5 and 6 the cores of new TVS that move perpendicularly to the original direction are being formed.

4. Conclusion

As a result of mathematical modeling we demonstrate the stages of interaction of two stable toroidal vortex structures. The interaction scenarios depend on masses and initial velocities of vortices and may look as follows:

a) Absorption of a vortex by another one and further movement of the joint vortex in the direction defined by the stronger initial vortex.

b) Disintegration of both vortices and formation of new structures moving perpendicularly to the initial motion.

Our mathematical model allows to distinguish the details of described interaction stages. This model makes it possible to conclude what processes occur in a real collision of vortex structures and to predict the result of the collision.

References

[1] Yusupaliev U., Savenkova N P, Skladchikov S A. et al. *Bulletin of the Lebedev Physics Institute* 2011. **38**. P. 275.

[2] Yusupaliev U., Savenkova N P, Skladchikov S A. et al. *Moscow university physics bulletin.* 2013. **68**. 4. P.317.

[3] Yusupaliev U, Vinke E E. , Shuteev C A. , Yusupaliev P U. *Bulletin of the Lebedev Physics Institute.* 2010. **37**. 8. P. 227.
[4] Anpilov S V, Bychkov V L, Savenkova N P, Skladchikov S A 9th International Conference on Plasma Assisted Technologies (ICPAT-9). Saint-Petersburg, Russia, 2014. **1**. P. 79.
[5] Bychkov V L., Anpilov S V., Savenkova N P. *Russian Journal of Physical Chemistry B*. — 2014. **8.** 1. P.50.
[6] Laponin V S, Savenkova N P. *Computational Mathematics and Modeling*. 2014. **25**. 1. P.1.

The stability issues in problems of mathematical modeling

Mokin A. Yu., Savenkova N. P., Udovichenko N. S.

Moscow State University, Faculty of Computational Mathematics and Cybernetics, Russia

E-mail: nudovichenko@mail.ru, mkandrew@mail.ru

Abstract. In the paper it is briefly considered various aspects of stability concepts, which are used in physics, mathematics and numerical methods of solution. The interrelation between these concepts is described, the questions of preliminary stability research before the numerical solution of the problem and the correctness of the mathematical statement of the physical problem are discussed. Examples of concrete mathematical statements of individual physical problems are given: a nonlocal problem for the heat equation, the Korteweg-de Fries equation with boundary conditions at infinity, the sine-Gordon equation, the problem of propagation of femtosecond light pulses in an area with a cubic nonlinearity.

1. Introduction

The stability concept is multifaceted. The concept of physical stability differs from the stability definition in mathematics. But physicists and mathematicians try to solve one problem together - the problem of adequate mathematical modeling. To obtain a correct solution, it is very important to study in detail the stability at different stages of solving the mathematical modeling problem: in a physical experiment, in a mathematical formulation of the problem, in a difference scheme, i.e. in a numerical method for the problem solving.

The interpretation of the stability concept in physics is given in S.I. Ozhegov's dictionary [1]: "Stable." Standing, holding firmly, without hesitation, not falling. "Stable prop." Stable equilibrium (restored after a slight deviation.) "

Thus, from the physical point of view, we can talk about stability, if, with the introduction of small deviations in the system, they will have little effect on the behavior dynamics. See, for example, the problem of oscillations [2]. Sometimes physical stability is essentially technological stability. That is, the physical process has a property: with a random effect of some perturbation on this process, it spontaneously returns to its original state. The stability problems of technological processes are singled out in a separate series. A vivid example of technological stability is the problem of the MHD stability of an aluminum cell [3]. Stable process of electrolysis has technological stability.

Usually, the mathematical formulation of the problem of a physical experiment modeling is a system of ODEs (ordinary differential equations) or a system of non-stationary partial differential equations. Here the question arises of the correspondence between the numerical solution and the result of the physical experiment. Moreover, it is quite often in the construction of a mathematical model, some data that are considered insignificant can be ignored.

There are two questions: the correspondence of the obtained numerical solution to the unknown exact solution and the correspondence of the unknown exact solution to the observed result of the physical experiment. In the 20th century, many theorems on the stability of various problems were developed, which were based on the stability theory of A.M. Lyapunov [4]. This theory was developed to study the dynamics of the solution of ODE systems in the second half of the 19th century. For example, there is the Arnold-Moser theorem on the stability of the equilibrium positions of a Hamiltonian system with two degrees of freedom in the general elliptic case [5, 6].

The basis of A.M. Lyapunov's theory is precisely definition of physical stability. This the stability of a differential problem definition is closely connected with the notion of the problem correctness [7]:

1) the solution exists;
2) the solution is unique;
3) for small perturbations of the boundary conditions the solution also varies insignificantly.

With the existence and uniqueness of the differential problem solution, we can speak about its correctness and obtain a numerical solution. That is, we can build a difference scheme (set a difference problem). As a rule, mathematical statement of spatial homogeneous models reduces to a numerical solution of systems of differential equations, since an analytical solution can be written out only for an extremely narrow class of problems. If the difference scheme is stable, we obtain convergence to the solution of the differential problem. If difference scheme is unstable, there is no convergence: it is necessary to choose another difference scheme. We note that a class of ill-posed problems [8] exists. The theory for such problems solution is currently developing rapidly.

The basis of the numerical solution of the problem lies in the transition from continuous data to discrete data. We introduce a difference scheme with step it in time and hover space in the case of a spatially-inhomogeneous problem. It should be noted that the numerical differentiation is incorrect from the machine point of view. The reason is the limitation of the computer mantissa: when we calculate the difference derivative by the formula $\frac{\Delta y}{\Delta t}$ at $\Delta t \to 0$, we obtain division by zero and the result is infinity.

Nevertheless, in practice, conditions are possible with a finite Δt and numerical solution that coincides with the specified accuracy with the solution of the differential problem. But it is necessary, as mentioned above, to consider preliminary the questions of the correctness of the problem mathematical formulation, and also the stability of the chosen numerical method for its solution.

There are different approaches to the problem of finding the solution numerically.

2. Analytical solution

Of course, solving non-stationary equations and finding a solution as a function of time, we can investigate its behavior with increasing t.

But first, it is often not possible to obtain an analytical solution in general form. And secondly, it is rather difficult to analyze the already written out solution. Therefore, such approach is extremely rare.

3. Numerical methods

The method should also be stable. The application of the method stability to physical stability is carried out. The physical stability of such problems is determined experimentally. That is, when it is impossible to obtain an analytical solution, the operator (linear, nonlinear, self–conjugate, non–self–adjoint) is discretized, the solution is solved numerically (an experiment is carried out) – and the solution behavior is examined on physical experiment. The solution of the problem by a numerical method must ensure convergence to the solution of the differential problem at the grid nodes.

4. Special methods that move away from the classical, according to K.I. Babenko

Methods K.I. Babenko – methods without saturation [9], i.e. the smoothness of the solution is used. Ordinarily, the smoothness of the function is not used. However, one differential equation can be taken of different orders for the difference scheme. By the methods of K.I. Babenko it is recognized about the smoothness of the solution experimentally, and these data are used to construct the scheme.

5. Method of separation of variables with subsequent study of the spectrum

The method of separation of space-time variables is one of the most popular methods of solving (linear problems). As a result of which we obtain an eigenvalue problem

$$X(x,t) = A(x) * e^{\lambda t}.$$

The analysis of the operator spectrum more easily and visually characterizes the dynamics of the process. Depending on the properties of the spectrum, we can conclude that the initial stationary process is stable or not. In this case, the instability of the physical process is characterized by an unlimited increase in the characteristics of the given process, for example, the amplitude of the pendulum oscillations. This is clearly seen in physical experiments.

There is a whole direction of stability research using the properties of the spatial operator spectrum. Based on the theory of A.A. Samarskii for operator problems[10], the spectral theory of stability research was developed by A.V. Gulin [11]. It is based on the spectral properties of the basic operator of the difference scheme.

According to the theory of A.V. Gulin a difference problem must be stable too. Its stability, as well as the differential analog, is also characterized by the properties of the spectrum, the solution of the discrete eigenvalue problem. Moreover, all the properties of the difference problem, including the spectral ones, must coincide with an accuracy of the approximation order for the corresponding differential problem.

The application of the spectral theory is shown on the example of a nonlocal problem for the heat equation in [12]. For a differential problem with a nonlocal boundary condition

$$\frac{\partial u}{\partial t} = \frac{\partial^2 u}{\partial x^2}, \quad 0 < x < 1, \quad t > 0,$$
$$u(x,0) = u_0(x), \quad 0 < x < 1,$$
$$u(0,t) = 0, \quad \gamma \frac{\partial u}{\partial x}(0,t) = \frac{\partial u}{\partial x}(1,t),$$

where γ is given numerical parameter, the stability of the difference scheme with weights

$$\frac{y_i^{n+1} - y_i^n}{\tau} = \sigma y_{\bar{x}x,i}^{n+1} + (1-\sigma) y_{\bar{x}x,i}^n, \quad i = 1,2,\dots,N-1,$$
$$\frac{y_N^{n+1} - y_N^n}{\tau} = \frac{2}{h}\left[\sigma\left(\gamma y_{x,0}^{n+1} - y_{\bar{x},N}^{n+1}\right) + (1-\sigma)\left(\gamma y_{x,0}^n - y_{\bar{x},N}^n\right)\right],$$
$$y_i^0 = u_0(x_i), \quad i = 0,1,\dots,N, \quad y_0^{n+1} = 0, \quad n = 0,1,\dots$$

is studied on the basis of the spectral properties of second-order non-self-adjoint difference operators with a parameter γ, appearing in nonlocal boundary conditions. Stability criteria are obtained in the space with a special energy norm that does not increase on the solution of the difference problem and is equivalent to the standard mean-square norm.

The differential and difference problems can be linear and nonlinear by the properties of the operator. The spectral theory works well for linear eigenvalue problems, whose solution can form a discrete spectrum, a continuous spectrum, or a piecewise- continuous spectrum. But we have to study the stability of nonlinear schemes. Then we obtain a nonlinear nonstationary eigenvalue problem. The study of nonlinear problems has been carried out for a long time and is really relevant at the present time.

When studying the stability of a nonlinear problem, the concept of orbital stability is often used. This concept of stability was first introduced by A. Poincaré [13], based on the theory of A.M. Lyapunov. We consider not a concrete solution, but a phase space, the points of which characterize the state of the dynamic or physical model as a whole, the change of this state with the passage of time. Orbital stability describes the behavior of a closed trajectory (orbit) under the action of small external perturbations.

IOP Conf. Series: Journal of Physics: Conf. Series **996** (2018) 012005 doi:10.1088/1742-6596/996/1/012005

A bright example of a nonlinear problem is the problem of studying waves and oscillations. Waves of small amplitude are described by linear differential equations, which are studied. But if the amplitude is not small, the differential equation becomes nonlinear. For example, the Korteweg-de Fries equation (KDF) [14]

$$\frac{du}{dt}+6u\frac{du}{dx}+\frac{d^3u}{dx^3}=0,$$

which describes waves in shallow water. In the form of a traveling wave $u(x,t)=f(x-ct)$ the KDF equation can be solved analytically. If $u(x,t)\xrightarrow[x\to\infty]{}0,$ then one of the exact solutions has the

$$u_1(x,t)=\frac{c}{2}sch^2\left(\frac{\sqrt{c}}{2}(x-ct+\sigma)\right),$$

where σ - integration constant. The solution $u(x,t)$ - is a solitary wave localized in a small region. KDF has an infinite number of exact solutions $u_n(x,t)$, that contain 2n arbitrary parameters c_i and σ_i and also behave as a superposition of independent isolated waves of the form $u_1(x,t)$ for $t \ll 0$ or $t \gg 0$. These waves can catch up with each other or collide with each other for a finite period of time, but then they always return to their original state (although a phase shift is possible). They do not annihilate, they behave like particles. The solution $u_n(x,t)$, that consists of n-isolated waves is called n–soliton solution.

If the differential equation is linear, then the superposition principle holds for it, that is, if $u_i(x,t)$, $i=1,...,N$ is a particular solution, then $\sum_{i=1}^{N}\alpha_i u_i$ is also a solution. For nonlinear equations this principle is not valid, but the existence of solutions $u_n(x,t)$ is an analogue of the superposition principle. KDF can be reduced to a system that has infinitely many degrees of freedom, that is, it is an absolutely integrable system and has infinitely many first integrals (invariants).

In addition to KDF, the Kadomtsev-Petviashvili (KP) equations, Toda equations and other soliton equations have the property of the superposition principle. We emphasize that these equations also have ordinary nonlinear solutions, apart from the isolated soliton solutions. Soliton solutions are stable by definition.

There are methods for finding soliton solutions: the inverse scattering method [15], the Hirota method (that is, the bilinear method), the theory of quasi-periodic solutions, etc., and also the specially developed methods M1 and M2 [16] and the method for constructing difference schemes for nonstationary equations [17], which are based on finding the invariants of equations systems.

Numerical solutions of methods must converge to soliton solutions of different modes, and the convergence depends little on the form of the initial function. For example, in the paper [18] we can see the application of the iterative method M2 for finding soliton solutions of various nonlinear problems:

for the KDF equation with boundary conditions at infinity

$$\frac{\partial u(x,t)}{\partial t}+6u(x,t)\frac{\partial u(x,t)}{\partial x}+\frac{\partial^3 u(x,t)}{\partial x^3}=0,\ x\in R,\ t>0,$$

$$u(\pm\infty,t)=0,\ u'(\pm\infty,t)=0,\ u''(\pm\infty,t)=0,\ t>0,$$

$$u(x,0)=u_0(x),\ x\in R,$$

for the sine–Gordon equation of the form

IOP Conf. Series: Journal of Physics: Conf. Series **996** (2018) 012005 doi:10.1088/1742-6596/996/1/012005

$$\frac{\partial u^2(x,t)}{\partial t^2} = a\frac{\partial^2 u(x,t)}{\partial x^2} + b\sin(cu(x,t)),$$
$$u_x^{'}(\pm\infty,t) = 0, \quad u_t^{'}(\pm\infty,t) = 0,$$

where *a, b, c* are given;

for the problem of femtosecond light pulses propagating in a cubic nonlinear medium [19]

$$\frac{\partial u}{\partial z} + i\frac{\partial^2 u}{\partial t^2} + i\alpha\frac{\partial}{\partial t}(|u|^2 u) = 0, \quad 0 < t < L, \quad z > 0,$$
$$u(z,0) = u(z,L) = 0,$$
$$u(0,t) = u_0(t),$$

where $u(z,t)$ is the complex amplitude of a pulse propagating along the z axis (measured in units of the dispersion spreading length) normalized to the maximum value, t is the time in the accompanying coordinate system normalized to the duration of the main pulse,

L is the dimensionless time interval, α is the ratio of the initial pulse power to the characteristic self-action power, γ is the coefficient characterizing the rate of change of the nonlinear polarization.

Thus, the study of the problem stability at all stages of its solution ensures the correctness of the solution, i.e. the correlation of the solution of the differential problem with the solution of the difference problem, and also the obtaining of a model whose solution corresponds to the results of a physical experiment. All of the above provides a qualitative correspondence of the conclusion about the stability of the physical process on the basis of a numerical solution. In order to answer the question of the quantitative correspondence of the characteristics of the physical process stability (growth increments, damping increments, etc.), the theory of accuracy research based on calculation on condensing grids has been developed in mathematics [20].

References

[1] Ozhegov S I, *Dictionary of the Russian language*. The 22nd edition. 1990.
[2] Kollatts L, *Tasks for eigenvalues with technical applications*. M.: Nauka. 1968.
[3] Davidson P A, Lindsay R I, *Journal of Fluid Mechanics*, 1998. **5**. P.273.
[4] Lyapunov A M, *The general problem of the stability of motion.* Kharkov: Publishing house of the Kharkov Mathematical Society. 1892.
[5] Arnold V I, *DAN SSSR*. 1961. **137**. 2. P.255.
[6] Moser Yu. *Lectures on Hamiltonian systems*. M.: MIR 1973.
[7] Samarskii A A., Gulin A V. *Stability of difference schemes*. M.: Nauka. 1973.
[8] Tikhonov A N., Denisov A M. *Inverse and ill–posed problems. Training course:* Lomonosov Moscow State University. Since 1966.
[9] Babenko K I, *DAN SSSR*. 1978. **241**. 3. P.505.
[10] Samarskiy A A, *The theory of difference schemes.* 3rd ed. M. :Nauka, 1989.
[11] Gulin A V. *Symmetrizable difference schemes*. M.: MAKS-Press. 2004.
[12] Gulin A V., Udovichenko N S. *Stability of a nonlocal difference problem with a complex parameter. Differential equations.* M .: Nauka. **47**. 8. P.1105.
[13] Poincare A. Selected Works. Moscow: Nauka. **1-3**, P.1971.
[14] Miva T., Jimbo M., Date E. *Solitons: differential equations, symmetries and infinite-dimensional algebras.* Moscow: MTsNMO. 2005.
[15] Zakharov V E, Shabat A B. *Funkts. analysis and its adj.* 1979. P.13–22.
[16] Savenkova N P, Laponin V S. *Bulletin of the Siberian State Aerospace University named after Academician MF Reshetnev*. 2013. **2(48).** P.81.
[17] Trofimov V.A. *Journal of Computational Mathematics and Mathematical Physics*. 1998. **38:12**. P.2055.

[18] Savenkova N P, Laponin V S. *Bulletin of Moskovskogo Universiteta. Series 15: Computational Mathematics and Cybernetics*. 2013. **2**. P.5.

[19] Dorokhova T V., Savenkova N P., Trofimov V A. *Applied Mathematics and Informatics.* 1999. **2**. P.63.

[20] Marchuk G I, Shaidurov V V. *Improving the accuracy of solutions of difference schemes.* M.: Nauka. 1979.

IOP Conf. Series: Journal of Physics: Conf. Series **996** (2018) 012006 doi:10.1088/1742-6596/996/1/012006

Analytical properties of nonequilibrium threshold in shock waves

M M Kuznetsov, Y D Kuleshova, Yu G Reshetnikova and L V Smotrova

Moscow Region State University, Radio street 10A, Moscow, 105005, Russia

E-mail: kuznets-omn@yandex.ru, juliaybogdanova@mail.ru, gau1972@mail.ru, lilysmotrova@mail.ru

Abstract. Are the main results of numerical research of the effect of high speed threshold Nonequilibrium activated kinetic processes in shock-compressed gases. In parallel to it are corresponding analytical results of the authors.

1. Introduction

Analytical properties of threshold of kinetic processes in the shock waves are directly related to the so-called effect of high-speed translational Nonequilibrium [1]. The effect of significant excess number of pairs of molecules that have overcome the threshold energy of a chemical reaction inside the shock wave front, over the appropriate number of pairs of molecules behind it, was previously installed in the paper [2] on the through numerical Monte Carlo simulation.

The numerical research of shock wave Monte Carlo also identified all of the basic physical factors accelerating the flow of kinetic processes in shock waves:

— effectively reduction the threshold of chemical reactions inside the front shock wave due to the nature of the bimodal distribution function Tamm-Mott-Smith;

— speed reduction of equilibrium chemical reactions in a "hot" area behind the front shock wave due to the strong dilution of " Rayleigh gas " by predominant light carrier;

— reducing the speed of the high-speed equilibrium of chemical reactions in a "hot" area behind the front shock wave due to energy dissociation expense (reduction of equilibrium statistical temperature compared to kinetic inside front shock wave);

— velocity acceleration high-speed chemical reactions due to anisotropy field of kinetic temperatures inside the shock wave.

It should also be noted that the relative effect of a high-speed "overshoot" increases with the energy threshold of inelastic collisions between pairs of molecules. For threshold velocity constants of chemical reactions this fact repeatedly emphasized in works of V Yu Velikodniy (see for example [3]). In the works of the same author, apparently, was first proposed by an analytical approach to the study of threshold progressively nonequilibrium processes on the basis of the bimodal distribution of molecules within the shock wave. His research focused on the impact of the first factor to the effect of "overshoot" speeds of translational nonequilibrium threshold chemical reactions. Influence of anisotropy field of kinetic temperatures on the magnitude of this effect was considered in [1].

An analytical approach to the study of the effect of threshold progressively Nonequilibrium in shock wave has some fundamental advantages (in the sense of final results) compared to numerical.

But this is achieved through a substantial simplification of the mathematical model of the consider effect. So, using the ellipsoidal distribution function of the molecules in the gas mixture components [1], you can show a certain interdependence of four determining physical factors. To take into account the impact of these factors on translational nonequilibrium velocity of threshold chemical reactions can offer a universal formula for some chemical kinetics [4].

2. Binary gas mixtures with constant and variable approximation parameters in Maxwellian "wings" of Bimodal distributions

Previously two theorems on high velocity translational nonequilibrium in shock wave for one-component (or pure) gas were formulated in authors works. Besides was noted that the proof of this theorems in the case of a shock-compressed gas binary mixture involves overcoming some significant difficulties. Chief among that is the fact that in gas mixtures the bimodal molecular velocity distribution cannot be unconditionally applied for the calculation of shock wave structure, as has been done in a simple gas. As a rule, the range of applicability of the classical bimodal Tamm – Mott-Smith distribution with constant approximation of auxiliary macroscopic velocities and kinetic temperatures for groups of molecules in supersonic and subsonic "wings" of this distribution is limited by small concentration values of one of the mixture components. In this case only it is possible to preserve all the advantages of the bimodal approximation of partial distribution functions in binary gas mixture to obtain a simple analytical solution for shock wave structure. Because of that the obtaining of necessary and sufficient high velocity translational nonequlibrium conditions for arbitrary values component concentration of gas mixtures becomes much more complex. In this paper a rigorous formulation and the substantiation of these conditions is given. These conditions apply to the case of variability of the auxiliary approximation macro parameters of "hot" bimodal distribution wing in binary gas mixture. All the same time the values of auxiliary macroscopic velocities and kinetic temperatures for groups of molecules in subsonic "wing" are constant. In binary mixtures of gases with distribution functions for the light and heavy components can be three distribution functions of the module of relative molecular velocity for pairs molecules. These functions are: G(ll) is the distribution function of pairs inside a light-light component, the second – G(lh) – function pairs light-heavy component, the third – G(hh) is a function of pair heavy-heavy component. The numerical realization of corresponding overshoot effect of bimodal distributions for pair molecules was examined in particular case of small concentration of heavy component in binary gas mixture (Rayleigh gas). It is known that in Rayleigh gas this effect for function G(hh) is most noticeable. The numerical calculation of this effect took into account the molecular rotational degrees of freedom in polyatomic gas.

3. Analytical estimate of high-speed overshoot value for pair molecules distribution in binary gas mixture

Analytical properties of bimodal distribution of pairs molecules allow to establish the existence of maximum values for the relative effect of "overshoot" in the number of pairs of molecules inside front shock waves compared to the appropriate number of pairs of molecules behind the shock wave. Furthermore, to estimate the magnitude of the "overshoot" of pairs of molecules can offer universal analytical expression, finding which requires no detailed numerical calculation of shock wave front structure. For a simple gas it was done in [1]. For binary gas mixtures the presence of maximum for the value of the corresponding "overshoot" was found in [5]. In the present work a formula for its estimate is provided:

$$\left[\left(G^{(neq)}/G_1^{(eq)}\right)-1\right] > g_{0,1}^{(\alpha,\beta)} \cdot g_{1,0}^{(\alpha,\beta)}\left[g_{0,1}^{(\alpha,\beta)} + g_{1,0}^{(\alpha,\beta)} + g_1^{(\alpha,\beta)}\right]^{-1} > 0$$

Here's the first set of square brackets is the value of corresponding of "overshoot" effect for distribution functions of pairs molecules inside the wave front - $G^{(neq)}$ belong to the corresponding translational equilibrium value - $G_1^{(eq)}$. Other designations are given in [5].

4. Conclusion

Note that all values on to the right of the first inequality, enables to estimate the relative effect of "overshoot" magnitude does not require for its computing knowledge structures of front shock wave. The structure of shock wave knowledge allows to replace the first inequality to equality. In this equality compared to inequality appears an additional multiplier order the unit. However, for practical applications, effect of "overshoot", as shows experiment [6], shall be not less than 10^4-10^6. Knowledge factor order 1 is not fundamental.

5. Acknowledgments

This study is supported by Russian Foundation for Basic Researches, grant 17-07-00945A

References

[1] Kuznetsov M M, Kuleshova Ju D and Smotrova L V, *Bulletin of the Moskow State Regional University* 2012. (2) P.108
[2] Genich A P, Kulikov S V, Manelis G B, Chereshnev S L, *Izv. ANSSSR MZhG* 1990. (2) P.144
[3] Velikodniy V Yu, 1990 *Molecular dynamics of gas and the mechanics of nonuniform media* M.: Nauka 1990 P.41
[4] Kuznetsov M M and Smotrova L V, *Bulletin of the Moskow State Regional University* 2013. (3) P. 66
[5] Kuznetsov M M, Kuleshova Ju D, Smotrova L V and Reshetnikova Yu G, *Bulletin of the Moskow State Regional University* 2016. (3) P. 84
[6] Genich A P, Kulikov S V, Soloviova M E, Chereshnev S L, *Fundamental Problems of Shock Vaves Fhysics.* Book of Abstracts 1987. **1** (2). P. 240

The H-theorem and equation of state for kinetic model of imperfect gas

A. M. Bishaev[1], V. A. Rikov[2], M. V. Abgaryan[3]

[1]MIPT
[2]CC RAS
[3]RIAME MAI

E-mail: bishaev@bk.ru, chatebrian@mail.ru

Abstract. In the offered article, having used earlier constructed kinetic model for imperfect gas, the equation of state for such gas which takes place which is able in a thermodynamic equilibrium is received and also expression for critical temperature as functions is received from an interaction potential between molecules.

1. Introduction

It is known that N.N. Bogolyubov, applying asymptotic methods, received Boltzmann's equation from the BBGKY system (a conclusion can be found in the section ADDITION of the monograph [1]). In this method Boltzmann's equation turns out as the equation for definition of the one-particle function of distribution in λ- scale, the BBGKY system which is a term of an order zero at decomposition in small parameter $\varepsilon = n_0 d^3$. It is accepted above that n_0- the reference density of molecules of gas, $d \approx 10^{-8}$ sm, is molecular size (or atom). Usually take the size on which the potential of an intermolecular interaction is other than zero for this size, $\lambda = \frac{1}{n_0 d^2}$ is a path length of molecules in gas.

When a conclusion of the equation of Boltzmann is considered, always the potential with which molecules interact among themselves is considered the given. And it is supposed that molecules of gas interact as the pointwise centers of a repulsion. In reality, as we know, the interaction potential of molecules has both the repellent, and attracting components therefore Boltzmann's equation is fair for an ideal dilute gas (see [2]), i.e. for gas in which it is possible to neglect an average potential energy of an attraction of molecules in comparison with their average kinetic energy.

There are many works in which try to consider in Boltzmann's equation of the amendment of an order ε. The equations which are turning out in such procedure are treated as the kinetic equations for the dense gas. Already in [1] the procedure of accounting of effect of a shadowing of molecules is offered. The serial asymptotic procedure of accounting of terms of an order ε is bound to consideration of collisions of three and more numbers of molecules at the same time that leads to emergence in the made mention procedure of the dispersing integrals. In [3] this problem it is solved by means of use of the principle of easing of correlations at expression for multiparticle cumulative distribution functions.

It should be noted that injected parameter ε less unit practically for all environments is higher (this fact is available in [3]). For example, for iron (Fe) whose density - $\rho = 7,8\frac{2}{cм^3}$ $\varepsilon \cong 0.08$ more, than is ten times less than unit (for water $\varepsilon \cong 0.02$). From this it is possible to draw a conclusion that the direction connected taking into account only effects of consolidation of gas will hardly lead to creation of the kinetic equations describing behavior of actual environments.

There is rather large number of works where try to consider not ideality of gas in Boltzmann's equation (the attracting component in the potential of interaction of molecules). In [3] rather complete review of these works is given.

Noticeable contribution to comprehensions of problems, the bound to creation of the kinetic theory of imperfect gas were brought by Yu.L. Klimontovich. By it (see [4]) one more parameter of length characterizing the environment was injected. It is $l = n_0^{-1/3}$. It is treated as distance between molecules. Then, from above told follows that $\frac{d}{l} = \varepsilon^{1/3}$, and size, inverse is equal to Knudsen number which faces a collision integral in Boltzmann's equation $l/\lambda = \varepsilon^{2/3}$. From this it follows that the collision integral is nonanalytic function of multiparticle cumulative distribution functions therefore its decomposition on the whole degrees does not allow to consider correctly the effects bound to increase in density of the environment. Yu.L. Klimontovich in [4] also specified that for the kinetic description of imperfect gas it is necessary to use a two-particle cumulative distribution function that, generally speaking, is clear because the average potential energy of molecules will be expressed through a two-particle cumulative distribution function. It in [5] constructed the equation for the two-particle cumulative distribution function describing imperfect gas. It had no noticeable resonance. Probably because from this equation it was impossible to pass to ideal gas.

Absolutely other approach to a problem was offered by A.A. Vlasov ([6]). He entered a concept of capacity of the self-consistent field and constructed the kinetic equation for evolution of the cumulative distribution function which is in this field. In [3] data on area of applicability of approach of A.A. Vlasov are provided and also it is specified that now representations of a collision integral as the sums of boltsmanovsky integral, vlasovsky and still any terms are used.

The most relief effect of action of attractive forces between molecules is formations of the bound states those molecules which kinetic energy does not allow to overcome action of attractive forces. If to trust the author [3], then the problem of accounting of the bound states is still far from the decision, and the available kinetic equations considering the specified effect are too difficult for the analytical and numerical research.

In [7], the kinetic model considering effect of formation of the bound states was constructed. In this work the equilibrium equation of state received on the basis of this model is investigated.

2. The kinetic equation for imperfect gas

In [7] the kinetic model for imperfect gas was constructed. Imperfect gas authors called gas in which it is impossible to neglect a potential energy of interaction of molecules in comparison with their kinetic energy. In this case in a phase space $\Gamma = \Omega \times D$ where Ω - high-speed space of the relative velocities of the molecules $\vec{g} = \vec{\xi}_1 - \vec{\xi}$ and D - the relative provisions - $\vec{r} = \vec{x}_1 - \vec{x}$ areas of the bound states which are described by a ratio $\frac{m}{4}\vec{g}^2 + U(r) \le 0$ where $U(r)$ is a potential of an intermolecular interaction are formed. In the assumption that the two-particle cumulative distribution function of the molecules which are in the bound state has a canonical distribution - $\bar{F}_d = f_0(t,\vec{q},\vec{w}) \cdot exp\left\{-\frac{1}{k\varphi(t,\vec{q},\vec{w})}\left(\frac{m}{4}\vec{g}^2 - U_0\left(\frac{d}{r}\right)^4\right)\right\}$ For such gas, the following system of the kinetic equations was received:

$$\frac{\partial f}{\partial t}+\xi^i\frac{\partial f}{\partial x^i}=J_B-J_{del},\ \frac{Ds}{Dt}=J_{del},\ \frac{Dh}{Dt}=\chi_1\int\limits_{g\le 2\chi_1}\frac{m}{4}g^2d\vec{g}\int\limits_0^{2\pi}\int\limits_0^{\pi}|\cos\omega|(\bar{f}-f_0)\sin\omega d\omega d\theta,$$
$$J_B=\int d\vec{\xi}_1(\int\limits_0^{b}\int\limits_0^{2\pi}g(ff_1'-ff_1)bdbd\theta),\ \frac{D}{Dt}=\frac{\partial}{\partial t}+w^i\frac{\partial}{\partial x^i}+a^j\frac{\partial}{\partial w^j},\ \vec{a}=-\frac{4\pi U_0 d^3 grad(n)}{3m},\ \chi_1=\sqrt{\frac{U_0}{m}}$$
$$\vec{w}=\frac{\vec{\xi}+\vec{\xi}_1}{2}$$
(1)

$$J_{del}=\chi_1 d^2\int\limits_{g\le 2\chi_1}d\vec{g}(\int\limits_0^{2\pi}\int\limits_0^{\pi}|\cos\omega|(\bar{f}-f)\sin\omega d\omega d\theta$$

B (1) , $f=f(t,\vec{x},\vec{\xi})$ - cumulative distribution function of molecules in the free state,

$$\bar{f}=f^2(t,\vec{\xi},\vec{x})=f^2(t,\vec{w},\xi,\vec{x});s=s(t,\vec{q},\vec{w})=\iint\limits_{\Omega_{del}}\bar{F}_d d\vec{r}d\vec{g},\ h=h(t,\vec{q},\vec{w})=\iint\limits_{\Omega_{del}}\left(\frac{mg^2}{4}+U(r)\right)\bar{F}_d d\vec{r}d\vec{g}\le 0$$

Ω_{del} - is an area of a phase space of molecules which are in the bound state; $s\Delta\vec{q}\Delta\vec{w}$ is a number of the molecules which are in the bound state whose coordinates and speeds are in the corresponding element of a phase space, and $h\Delta\vec{q}\Delta\vec{w}$ - their energy.

Through the macro parameters entered above function f,s,h are expressed as follows: $n(t,\vec{x})=\int fd\vec{\xi}+\int sd\vec{w}$ - numerical density gas; $n\vec{u}=\int\vec{\xi}fd\vec{\xi}+\int\vec{w}sd\vec{w}$. $\vec{u}$ is macroscopic speed of gas. $P^{ij}=m\int c_f^i c_f^j fd\vec{\xi}+m\int c_d^i c_d^j sd\vec{w}$, $\vec{c}_f=\vec{\xi}-\vec{u},\ \vec{c}_d=\vec{w}-\vec{u},\ i,j=1,2,3$ - stress tensor in imperfect gas. Kinetic temperature is defined as follows: $\frac{3}{2}nkT=\int m\frac{\vec{c}_f^2}{2}fd\vec{\xi}+\int(m\frac{\vec{c}_d^2}{2}s-\frac{1}{2}h)d\vec{w}$. Where k - Boltzmann constant. In [7] it is shown that from (1) all conservation laws follow.

$$\frac{\partial n}{\partial t}+\frac{\partial(nu_i)}{\partial x_i}=0,\ mn\frac{\partial u^j}{\partial t}+mnu^i\frac{\partial u^j}{\partial x_i}+\frac{\partial P^{ij}}{\partial x_i}-na_j=0,\ i,j=1,2,3,\ n_d u_d^i=\int w^i sd\vec{w},$$
$$\frac{3}{2}kn(\frac{\partial T}{\partial t}+u^i\frac{\partial T}{\partial x_i})+P^{ki}\frac{\partial u_k}{\partial x_i}+\frac{\partial Q_i}{\partial x_i}-n(u_d^i-u^i)a_i=0,\ \vec{Q}=\int m\vec{c}_f\frac{\vec{c}_f^2}{2}fd\vec{\xi}+\int\vec{c}_d(m\frac{\vec{c}_d^2}{2}s-\frac{1}{2}h)d\vec{w}$$

3. Equilibrium equation of state

As The concept of pressure for imperfect gas is entered also as well as for ideal - $3P=P^{11}+P^{22}+P^{33}=m(\int\vec{c}_v^2fd\xi+\int\vec{c}_d^2sd\mathbf{w})$. Then from determination of macroscopic sizes it is possible to receive the following ratio:

$$(P-\frac{1}{3}\int hd\vec{w})=nkT \tag{2}$$

The ratio (2) is a kinetic equation of state. In right members of the equations for s and h it is also possible to integrate on high-speed space and in case in quality of mezhdumolekulyarny potential the special case of potential of Sutherland was taken:

$$U(r)=\begin{cases}+\infty, & 0\le r<d\\ -U_0(\frac{d}{r})^4, & r\ge d\end{cases}$$

, to lead these equations to a look $\frac{Ds}{Dt}=J/3,\ \frac{Dh}{Dt}=-U_0J/5$, where $J=64\pi^2\chi_1 d^2(\frac{U_0}{m})^{3/2}(\bar{f}-f_0)$. Then $\frac{D}{Dt}(\frac{3}{5}U_0 s+h)=0\rightarrow\frac{3}{5}U_0 s+h=c=cont$. It is clear, that if the number of particles of s an element of volume of a phase space is equal in the bound state to zero, then their energy is equal in the bound state too to zero therefore s and h are always bound a ratio $\frac{3}{5}U_0 s(t,x,\vec{w}) + h(t,x,\vec{w}) = 0$ from where we have a ratio for definition: $\varphi(t,\vec{x},\vec{w}) = \varphi = const$ $\frac{h(t,\vec{x},\vec{w})}{s(t,\vec{x},\vec{w})} = -\frac{3}{5}U_0 = -\frac{3}{5}U_0\frac{Z_h(\bar{\chi})}{Z_s(\bar{\chi})}$, where $Z_s(\bar{\chi}), Z_h(\bar{\chi}), \bar{\chi} = \frac{U_0}{k\varphi}$ [7]. In case of the homogeneous relaxation $f = f(t,\vec{\xi})$ and $s = s(t,\vec{w})$. Respectively the system (1) will have the following appearance:

$$\frac{\partial f}{\partial t} = \frac{d^2}{2}\int d\xi_1 (\int_0^b\int_0^{2\pi} g(ff_1' - ff_1)bdbd\theta) - d^2\chi_1 \int_{g\le 2\chi_1} d\xi_1 \int_0^{2\pi}\int_0^{\pi} |\cos\omega|(\bar{f} - f_0)\sin\omega d\omega d\theta$$

$$\frac{\partial s}{\partial t} = d^2\chi_1 \int_{g\le 2\chi_1} d\mathbf{g} \int_0^{2\pi}\int_0^{\pi} |\cos\omega|(\bar{f} - f_0)\sin\omega d\omega d\theta, \; \frac{3}{5}U_0 s + h = 0 \tag{3}$$

$S(t,\vec{w}) = \int_{\Omega_{del}} \bar{F}_d \ln \bar{F}_d d\vec{r} d\vec{g}$. Then $H_d = \frac{k}{2}\int S(t,\vec{w})d\vec{w}$ is H- function of unit volume for molecules in the bound state. Considering representation for a cumulative distribution function of molecules in the bound state, we will receive that, in unit of time in unit volume for these we will define production H-functions as:

$$\dot{H}_d = \frac{k}{2}\int \frac{\partial S}{\partial t} d\mathbf{w} = \frac{kd^2}{2}\int(\int_{g\le 2\chi_1} (\int_0^{2\pi}\int_0^{\pi} (\ln f_0 + 1 + \frac{3U_0}{5k\varphi})(\bar{f} - f_0)R(\omega,\theta)d\omega d\theta d\vec{g})d\vec{w} \tag{4}$$

We will write down H-function of the molecules which are in an unbound state as usual:

$H_f = k\int\int f \ln f e^{1/2 + \frac{3U_0}{10k\varphi}} d\vec{\xi} d\xi_1$, then:

$$\dot{H}_d + \frac{\partial H_f}{\partial t} = kd^2/2\int\int\int_0^{\infty}\int_0^{2\pi} \ln\frac{ff_1}{f_1'f'}(f_1'f' - ff_1)bdbd\varepsilon d\vec{\xi}_1 d\vec{\xi} + \frac{k}{2}d^2\int\int_{g\le 2\chi_1}\int_0^{\pi}\int_0^{2\lambda} \ln\frac{f_0}{f}(\bar{f} - f_0)R(\omega,\theta)d\omega d\theta d\vec{g} d\vec{w} \le 0$$

(5)

Thus, for the kinetic model constructed above the H-theorem takes place. Then the solution of a task on the homogeneous relaxation at $t \to +\infty$ will come to an equilibrium state at which $\dot{H} = \dot{H}_d + \frac{\partial H_f}{\partial t} = 0$, from where follows that $f_1'f' = ff_1, \bar{f} = f_0$. Then $f = n_f(\frac{m}{2\pi kT_f})^{3/2}\exp\{-\frac{m}{2kT_f}\vec{\xi}^2\}$, n_f is density of the molecules which are in the free state, and T_f it is possible to treat as their temperature. As $\bar{f} = f_0$ then $f_0 = f^2(\vec{w}) = n_f^2(\frac{m}{2\pi kT_f})^3 exp\{-\frac{m}{kT_f}\vec{w}^2\}$. In case of a task about a relaxation of a conservation equation have the following appearance: $\frac{\partial n_f}{\partial t} + \frac{\partial n_d}{\partial t} = \frac{\partial n}{\partial t} = 0, \frac{\partial T}{\partial t} = 0 \to n = const, T = const$. Using equilibrium values of required functions, we will receive that $n_f + rn_f^2 = n$.

$T_f \frac{1 + \frac{rn_f\chi_f^{3/2}}{2}(1 + \frac{2}{5}\chi_f)}{1 + rn_f\chi_f^{3/2}} = T, \; r = 2\sqrt{\pi}d^3 Z_s(\bar{\chi}), \chi_f = \frac{U_0}{kT_f}$ - The kinetic equation of state (2) if to write down it through equilibrium values of cumulative distribution functions will take the following form: $(P + \frac{U_0}{5}rn_f^2\chi_f^{3/2}) = knT$.

Let's enter $V = \frac{1}{n_f}$. It makes sense of volume which occupies gas. The equation of state then will correspond as $(P + \frac{U_0}{5}\frac{r\chi_f^{3/2}}{V^2})\frac{V}{(1+\frac{r\chi_f^{3/2}}{V})} = kT$. Boltzmann's equation is a term of an order zero at decomposition of BBGKY of a line-up in small parameter $\varepsilon = nd^3$. In [1] it is shown that not ideality of gas has to be considered when. If to write down an equation of state, assuming that $\varepsilon\chi_f^{3/2} < 1$, then it will take a form $(P + \frac{U_0}{5}\frac{r\chi_f^{3/2}}{V^2})(V - r\chi_f^{3/2}) = kT$. It is easy to see that this Van der Waals equation in which phenomenological parameters a and b also have concrete expressions through macroparameters of gas and a constant of an intermolecular interaction. Applying the procedure of receiving expression to critical temperature which is given in [8], we will receive that $kT_{крит} = \frac{8}{135}U_0$.

4. Conclusion

The assumption that $\varepsilon\chi_f^{3/2} < 1$, corresponds to the fact that approach to a critical state happens from gas, at the same time critical temperature is expressed through the reference value of an intermolecular interaction potential of molecules. In this kinetic model also other interaction potentials between molecules can be used, then in expression for critical temperature there will be other constant before U_0. To receive an isotherm for imperfect gas at any values of temperature, it is necessary to receive (2), proceeding from the decision (3) (a task about a relaxation) that in real the moment is been working on.

This article work is performed at the Moscow Aviation Institute (MAI) with financial support of the Ministry of Education and Science of the Russian Federation within the agreement on granting a subsidy No. 14.577.21.0231 of 29.09.2016 between MAI and the ministry (the PNIER unique identifier - RFMEF157716X0231.)

References

[1] Chapmen S, Kauling T. «The mathematical theory of the nonuniform gases» Publishing house of foreign literature, Moscow,1960. 510 p.
[2] Kogan M. N. «Dynamics of rarefied gases». Prod. «Science», Moscow, 196. 440 p.
[3] Rudjak V. Y. «The statistical theory of dissipative processes in gases and liquids». Prod. «Science», Moscow, 1987. 270 p.
[4] Klimontovich U. L. «Statistical theory». Moscow, «Science». Main edition of physical and mathematical literature, 1982, 19 p.
[5] Klimontovich U. L. «Kinetic theory of imperfect gases and plasma». Prod. «Science», Moscow, 1975, 608 p.
[6] Vlasov A. A. «Nonlocal statistical mechanics». Prod. «Science», Moscow, 1976, 264 p
[7] Bishaev A. M., Rikov V. A. Creation of system of the kinetic equations for imperfect gas. Thermal physics of high temperatures, 2017, **55**. 1. P.31.
[8] D. V. Sivuhin Common course of physics, T. II. Thermodynamics and molecular physics. Prod. "Science", Moscow, 1975, 552 p

IOP Conf. Series: Journal of Physics: Conf. Series **996** (2018) 012008 doi:10.1088/1742-6596/996/1/012008

On interaction of P-waves with one-dimensional photonic crystal consisting of weak conducting matter and transparent dielectric layers

A A Yushkanov, N V Zverev

Moscow Region State University, 10a Radio st., 105005 Moscow, Russia

E-mail: zverev_nv@mail.ru

Abstract. An influence of quantum and spatial dispersion properties of the non-degenerate electron plasma on the interaction of electromagnetic P-waves with one-dimensional photonic crystal consisting of conductor with low carrier electron density and transparent dielectric matter, is studied numerically. It is shown that at the frequencies of order of the plasma frequency and at small widths of the conducting and dielectric layers of the photonic crystal, optical coefficients in the quantum non-degenerate plasma approach differ from the coefficients in the classical electron gas approach. And also, at these frequencies one observes a temperature dependence of the optical coefficients.

1. Introduction

At present, a study of dependence the characteristics of the conducting matter on the temperature and other external parameters represents a large interest [1]. It is motivated first of all by a development of the fine bio- and nanotechnologies. The conducting substances having visible temperature behavior of a number of parameters, are substances with low carrier electron number density having order 10^{-5} – 10^{-2} of atomic one and hence, they possess a weak conductivity. These substances are semimetals, semiconductors and their compounds. The conductance electron gas of the matters at such densities is almost non-degenerate. But at the same time, in case of little widths of these low conducting substances, the quantum properties of the conductivity electrons should be also taken into account.

In the paper, one studies numerically an interaction of the electromagnetic P-waves with one-dimensional photonic crystal. This crystal consists of alternating thin layers of weak conducting matter with low carrier electron density, and a transparent dielectric substance. One compares the optical power coefficients for the photonic crystal evaluated in the quantum non-degenerate electron plasma approach, with those in the classical electron gas theory. And also, the temperature dependence of the optical power coefficients is studied in the case of quantum approach.

2. The model

The photonic crystal has N flat layers of the low conducting matter with the same width d_1, and the $N-1$ flat layers of the transparent dielectric matter with the permittivity ε_2 having the similar width d_2 and separating the weak conducting layers. Such a photonic crystal is localized between two transparent dielectric media with permittivities ε_1 and ε_3.

Let the electromagnetic P-wave i.e. wave when the **E** vector lies in the incidence plane, is incident on the photonic crystal under the incidence angle θ from the medium with permittivity ε_1. Then the optical power coefficients, namely, the reflectance R, transmittance T_r and absorptance A, for the one-dimensional photonic crystal look as [2 – 4]:

$$R = \left|\frac{m_{21}}{m_{22}}\right|^2, \qquad T_r = \left|m_{11} - \frac{m_{12}m_{21}}{m_{22}}\right|^2 \operatorname{Re}\left(\sqrt{\frac{\varepsilon_3}{\varepsilon_1}}\frac{\cos\theta'}{\cos\theta}\right), \qquad A = 1 - R - T_r. \tag{1}$$

Here m_{jk} are the matrix elements of the transfer matrix for the photonic crystal:

$$\begin{pmatrix} m_{11} & m_{12} \\ m_{21} & m_{22} \end{pmatrix} = \mathbf{M}_{3out}^{-1}\left(\mathbf{M}_{2in}\mathbf{M}_{2pr}\mathbf{M}_{2out}^{-1}\right)^{N-1}\mathbf{M}_{1in}, \tag{2}$$

where the matrices $\mathbf{M}_{\alpha\, in}$, $\mathbf{M}_{\alpha\, out}$ and $\mathbf{M}_{2\, pr}$ in case of the P-wave look as ($\alpha = 1, 2, 3$):

$$\mathbf{M}_{\alpha\, in} = \begin{pmatrix} \cos\theta_\alpha - Z_P^{(1)}\sqrt{\varepsilon_\alpha} & -\cos\theta_\alpha - Z_P^{(1)}\sqrt{\varepsilon_\alpha} \\ -\cos\theta_\alpha + Z_P^{(2)}\sqrt{\varepsilon_\alpha} & \cos\theta_\alpha + Z_P^{(2)}\sqrt{\varepsilon_\alpha} \end{pmatrix}, \tag{3}$$

$$\mathbf{M}_{\alpha\, out} = \begin{pmatrix} \cos\theta_\alpha + Z_P^{(1)}\sqrt{\varepsilon_\alpha} & -\cos\theta_\alpha + Z_P^{(1)}\sqrt{\varepsilon_\alpha} \\ \cos\theta_\alpha + Z_P^{(2)}\sqrt{\varepsilon_\alpha} & -\cos\theta_\alpha + Z_P^{(2)}\sqrt{\varepsilon_\alpha} \end{pmatrix}, \tag{4}$$

$$\mathbf{M}_{2\, pr} = \begin{pmatrix} \exp(\mathrm{i}k_{2z}d_2) & 0 \\ 0 & \exp(-\mathrm{i}k_{2z}d_2) \end{pmatrix}. \tag{5}$$

In these equations, $\theta_1 = \theta$ is the angle of incidence, but θ_2 and $\theta_3 = \theta'$ are the refraction angles into the dielectric layers with ε_2 and into the medium with ε_3, respectively. The latter angles are evaluated according to relation

$$\sqrt{\varepsilon_1}\sin\theta = \sqrt{\varepsilon_2}\sin\theta_2 = \sqrt{\varepsilon_3}\sin\theta'. \tag{6}$$

Further $k_{2z} = \dfrac{\omega}{c}\sqrt{\varepsilon_2}\cos\theta_2$ is the projection of wave vector $\mathbf{k}_2$ in a dielectric layer with ε_2, onto the Z axis which is orthogonal to the layers surfaces of the photonic crystal and is directed towards the medium with ε_3. Here ω is the wave frequency and c is the vacuum speed of light.

And at the end, $Z_P^{(j)}$ ($j = 1, 2$) are the dimensionless surface impedances for the P-wave incident on the layer of the conducting matter. The impedances were evaluated in the paper [5] in the case of mirror reflection of the carrier electrons from the layer surfaces of the matter and look as follows:

$$Z_P^{(j)} = \frac{2\mathrm{i}c\omega}{d_1}\sum_n \frac{1}{k_n^2}\left(\frac{k_x^2}{\omega^2\varepsilon_l(\omega,k_n)} + \frac{(\pi n/d_1)^2}{\omega^2\varepsilon_{tr}(\omega,k_n) - (ck_n)^2}\right). \tag{7}$$

Here $k_n = \sqrt{\left(\dfrac{\pi n}{d_1}\right)^2 + k_x^2}$ and $k_x = \dfrac{\omega}{c}\sqrt{\varepsilon_1}\sin\theta$ is the projection of the wave vector $\mathbf{k}$ of incident wave onto the X axis which lies in the incidence plane and is parallel to the layer surfaces of the photonic crystal. Further, summation in (7) is performed over all odd numbers $n = \pm1, \pm3, \pm5,\ldots$ when $j = 1$, and over all even $n = 0, \pm2, \pm4,\ldots$ when $j = 2$. And finally, ε_l and ε_{tr} are the longitudinal and transverse dielectric functions of the electron plasma.

One has taken for investigation the dielectric functions of the quantum non-degenerate electron plasma with invariable collision frequency evaluated in [6, 7]:

$$\varepsilon_l^{(qn)}(\omega,k) = 1 - \frac{2}{Q^2}\frac{(\Omega+\mathrm{i}\gamma)G(\Omega+\mathrm{i}\gamma,Q)G(0,Q)}{\Omega G(0,Q) + \mathrm{i}\gamma G(\Omega+\mathrm{i}\gamma,Q)}, \tag{8}$$

$$\varepsilon_{tr}^{(qn)}(\omega,k) = 1 - \frac{1}{\Omega^2}\left(1 + \frac{\Omega G(\Omega+\mathrm{i}\gamma,Q) + \mathrm{i}\gamma G(0,Q)}{\Omega+\mathrm{i}\gamma}\right), \tag{9}$$

where

$$G(\Omega+\mathrm{i}\gamma,Q) = \frac{Q^2}{\sqrt{\pi}}\int_0^{+\infty}\frac{[(\Omega_+ + \mathrm{i}\gamma)(\Omega_- + \mathrm{i}\gamma) + (Q\xi)^2]\exp(-\xi^2)}{[(\Omega_+ + \mathrm{i}\gamma)^2 - (Q\xi)^2][(\Omega_- + \mathrm{i}\gamma)^2 - (Q\xi)^2]}\mathrm{d}\xi, \tag{10}$$

$$\Omega = \frac{\omega}{\omega_p}, \quad \Omega_\pm = \Omega \pm \frac{1}{2}\frac{\hbar}{m_e \omega_p} k^2, \quad \gamma = \frac{1}{\omega_p \tau}, \quad Q = k\sqrt{\frac{2k_B T}{m_e \omega_p^2}}. \tag{11}$$

Here ω_p is the frequency of the conductance electron plasma, τ is the electron relaxation time owing to collisions of the plasma electrons, m_e is the effective carrier electron mass, $\hbar$ is the Planck constant, k_B is the Boltzmann constant, and T is the temperature of electron plasma.

The dielectric functions (8) and (9) take into account both the spatial dispersion i.e. dependence of the functions on the wave number k, and the quantum wave properties of the electron plasma reflected in the terms containing the Planck constant.

The obtained results are compared with the data for the dielectric functions of the classical electron gas in the Drude – Lorentz approach [3, 5]:

$$\varepsilon_l^{(DL)}(\omega) = \varepsilon_{tr}^{(DL)}(\omega) = 1 - \frac{1}{\Omega(\Omega + i\gamma)}. \tag{12}$$

These dielectric functions of the classical electron gas are obtained from the functions (8) and (9) in the long-wave limit $k \to 0$. One sees that the functions reflect neither spatial dispersion nor quantum wave properties of the electron plasma.

3. Numerical results

To perform numerical calculation of the optical power coefficients for the photonic crystal with low conducting matter, one has to evaluate the plasma frequency and the electron relaxation time. These values are obtained according to the equations:

$$\omega_p = \sqrt{\frac{e^2 n_e}{\varepsilon_0 m_e}}, \qquad \tau = \frac{m_e}{e^2 n_e \rho_0},$$

where e is the elementary charge, ε_0 is the electric constant in the SI units, n_e is the carrier electron number density, ρ_0 is the static resistivity of the electron plasma. The latter two of them were taken from [8] for graphite when assuming the density n_e is 10^{-5} times the atomic one and the effective electron mass m_e is equal to the free electron one. Thus one gets $\omega_p = 5.67\cdot10^{13}$ s^{-1} and $\tau = 9\cdot10^{-12}$ s.

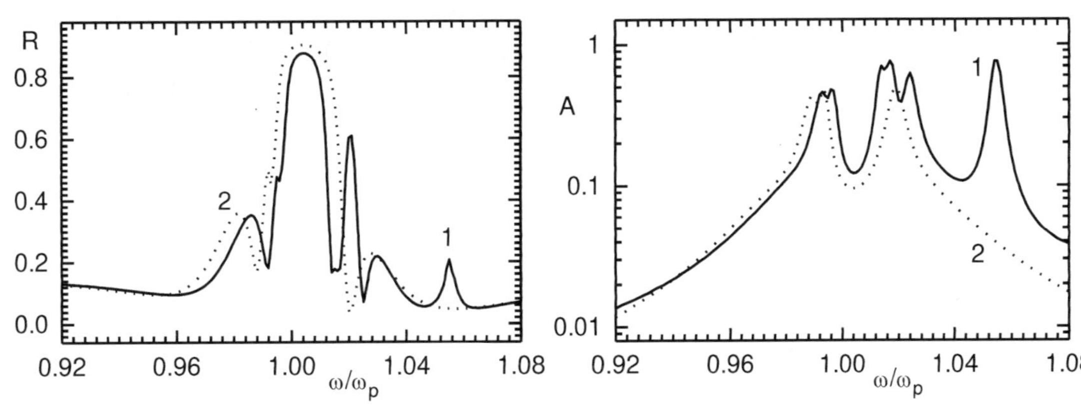

Figure 1. The reflectance R (left plot) and absorbance A (right plot) as functions of the frequency ω at $N = 6$, $\varepsilon_1 = 1$, $\varepsilon_2 = \varepsilon_3 = 2$, $\theta = 75°$, $d_1 = 100$ nm, $d_2 = 5$ μm, $T = 295$ K, $\omega_p = 5.67\cdot10^{13}$ s^{-1}: 1 – quantum non-degenerate electron plasma (solid line), 2 – classical electron gas (dotted line).

Using the values, one performs the numerical evaluations of the optical power coefficients according to the equations (1) – (12). Results of the calculations show first that for the frequencies $\omega \sim \omega_p$ when the conductor width is much less than the skin depth i.e. $d_1 << c/\omega_p$, and the dielectric layer width d_2 is of order of c/ω_p, the optical power coefficients in the quantum non-degenerate electron plasma approach with the dielectric functions (8) and (9) differ from the ones for the classical electron gas with functions (12) (see fig.1). So, an influence of both the spatial dispersion and quantum wave properties of the electron plasma on the optical power coefficients of the one-dimensional photonic crystal has been demonstrated. In comparison with the classical electron gas, additional resonant peaks of the power coefficients in the quantum non-degenerate plasma approach at the frequencies $\omega > \omega_p$ (fig.1) are caused by oscillations of the longitudinal plasmons [5] within the conducting layers of the photonic crystal that occurs just for the P-waves.

One observes also a temperature dependence of the optical coefficients at the frequencies $\omega \sim \omega_p$ for the quantum non-degenerate plasma (see fig.2). Such dependence has also resonant peaks due to the plasmon oscillations. One sees that the growth of the temperature leads to increase of the peak frequencies.

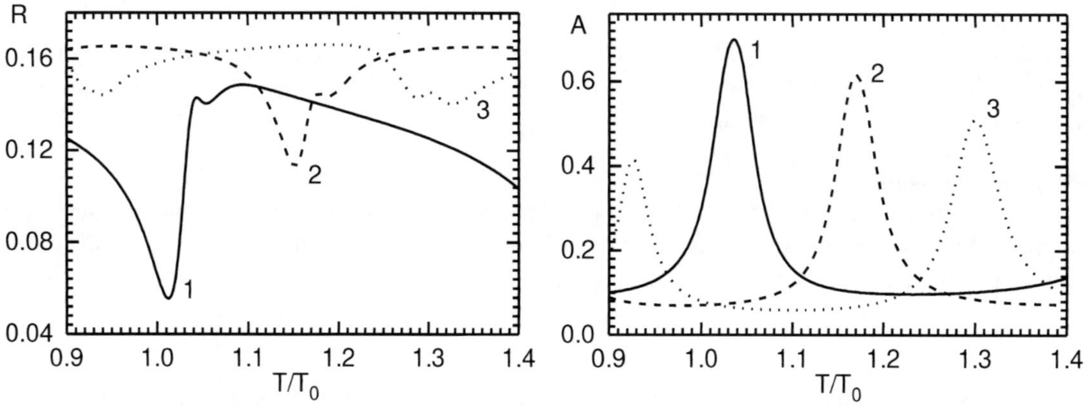

Figure 2. The reflectance R (left plot) and absorbance A (right plot) as functions of the temperature T for quantum non-degenerate electron plasma at $N = 6$, $\varepsilon_1 = 1$, $\varepsilon_2 = \varepsilon_3 = 2$, $\theta = 75°$, $d_1 = 200$ nm, $d_2 = 4$ µm, $T_0 = 295$ K, $\omega_p = 5.67 \cdot 10^{13}$ s^{-1}: 1 – $\omega = 1.07\omega_p$ (solid line), 2 – $\omega = 1.08\omega_p$ (dashed line), 3 – $\omega = 1.09\omega_p$ (dotted line).

The above effects of the quantum non-degenerate plasma on the power coefficients of the photonic crystal containing low conducting layers should be accounted for at the constructing and exploiting of fine optical devices having such photonic crystals.

4. Acknowledgements

The work is performed under partial support of the Russian Foundation of Basic Researches and Government of Moscow Region (project No. 17-47-500752).

References

[1] Faramarzi, Sh, Hervieuxa, P-A, and Bigota, J-Y, *J. of Optoel. and Advan. Mater.*, 2005. **7(6).** P.3083.
[2] Yeh, P, Optical waves in layered media. (NY, Wiley, 1988. 406 pp.)
[3] Gorelik, V S, and Kapaev, V V, *JETP*, 2016. **123(3).** P.373.
[4] Yushkanov, A A, and Zverev, N V, *Optics and Spectroscopy*, 2017. **122(2).** P.202.
[5] Jones, W E, Kliewer, K L, and Fuchs, R, *Phys. Rev.*, 1969. **178(3).** P.1201.
[6] Latyshev, A V, and Yushkanov, A A, *Theor. and Math. Phys.*, 2013. **175(1).** P.559.
[7] Latyshev, A V, and Yushkanov, A A, *Theor. and Math. Phys.*, 2014. **178(1).** P.130.
[8] Babichev, A P, Babushkina, N A, Bratkovskii, A M et al, Physical Quantities: Handbook / Ed. by Grigor'ev, I S, and Meilikhov, E Z. (In Russian. Moscow: Energoatomizdat, 1991. 1232 pp.)

IOP Conf. Series: Journal of Physics: Conf. Series **996** (2018) 012009 doi:10.1088/1742-6596/996/1/012009

On behavior peculiarity of electron plasma

N M Gordeeva [1] and A A Yushkanov[2]

[1]*Bauman Moscow State Technical University* 105005 Baumanskaya street, 5/1, Russia

[2]*Moscow Regional State University* 105005, Radio street,10a, Russia

e-mail: nmgordeeva@yandex.ru, yushkanov@inbox.ru

Abstract. The analysis of the analytical solution of the problem of the behavior of electron plasma in the AC electric field is fulfilled. Debye mode describes shielding of the external electric field in the plasma. The analysis of the region of existence of Debye mode, depending on the plasma parameters has been realized. A non-trivial dependence of the region of existence of Debye mode on the degree of degeneracy of the electron gas are revealed. For the case of nearly degenerate electron gas Debye mode has several areas of existence, depending on the frequency of the electric field.

1. Introduction

The plasma behavior near boundary are now intensely investigated [1,2]. On the shielding of the electric field close to the border Debye mode has a decisive influence. Debye mode is the solution of the kinetic Vlasov-Boltzmann and Maxwell equations. We will consider the kinetic equation in BGK approximation [3]. We will consider the case of small electric fields. This allows for the linearization of the kinetic equation. For the collisionless case, the solution on the behavior of plasma in one-dimensional case was first obtained by Landau. In this work it was firstly introduced the concept of collisionless damping (Landau damping) [4]. For collisional plasma the problem has been solved for degenerate case in [5]. The non degenerate case has been considered in [6].

2. The model and the solution

We will consider the one-dimensional case. So the electric field depends only on one coordinate. Designate this coordinate by x. According the assumptions, any solution can be represented as a decomposition of the field strength on their own decisions. In dimensionless form it looks like this [6]

$$e(x) = e_{as} + e_d(x) + e_c(x).$$

Terms e_{as}, e_d (x), e_c (x) are Drude, Debye, and van Kampen modes respectively. Note that the Drude mode does not depend on the x coordinate. However, the Debye mode is not always present in the decomposition for its solutions. It is only present when the dispersion function has a root at these parameters (frequency of the external electric field and the collision frequency in the plasma). Our goal was to determine the existence regions of Debye mode in the space of these parameters. To determine the existence of a root, the authors used the principle of the argument. This principle in the application to the conditions of this problem leads to next result. The complex plane was divided into regions, where exists and not exists root. The boundary of this region is determined by the parametric curve

$$\text{L:}\quad L(\alpha):\qquad \Omega = \sqrt{L_1(\mu,\alpha)},\qquad \varepsilon = \sqrt{L_2(\mu,\alpha)},\qquad 0 \le \mu \le +\infty,$$

$$L_1(\mu,\alpha) = \frac{s_0(\alpha)}{s_2(\alpha)} \frac{\mu^2\left[\lambda_0(\mu,\alpha)\left(1+\lambda_0(\mu,\alpha)\right)+s^2(\mu,\alpha)\right]^2}{\left(-\lambda_0(\mu,\alpha)\right)\left[\left(1+\lambda_0(\mu,\alpha)\right)^2+s^2(\mu,\alpha)\right]};$$

$$L_2(\mu,\alpha) = \frac{s_0(\alpha)}{s_2(\alpha)} \frac{\mu^2 s^2(\mu,\alpha)}{\left(-\lambda_0(\mu,\alpha)\right)\left[\left(1+\lambda_0(\mu,\alpha)\right)^2+s^2(\mu,\alpha)\right]}.$$

Here Ω is the dimensionless frequency of the external field, ε is the dimensionless collision frequency in the plasma, α is the dimensionless degeneration coefficient of the plasma. Numerical analysis of the presented expressions has been fulfilled.

Figure 1 presents parametric curves for several variants of partially degenerate plasma. The ratio of the degeneration of plasma α changes from -5 to 5. In the figure the following notation is used: D+ – region of existence of the root and D- is the area where there is no root. It is seen that graphs with α=-3 and α=-5 are practically identical. Also the graph for Maxwell plasma, where α=-∞, too, was close to 1 and 2.

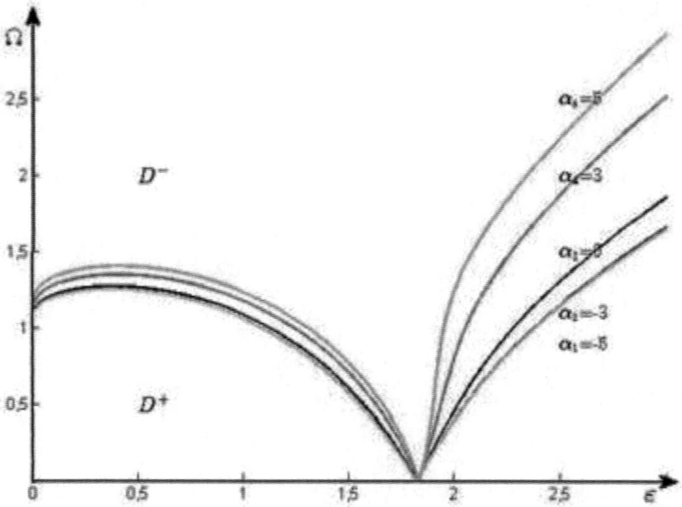

Figure 1. The dependence of the location of the curve L from the value of the coefficient α.

Further study of the behavior of the curve during the growth of α showed an interesting feature. The right branch of the curve, more curves and appear region in which the root (and therefore Debay mode) appears and then disappears. Figure 2 shows the curve L corresponding to the ratio of the degeneration of plasma α=30. It is seen that for ε=1,5 with increasing frequency of the external field we get the following picture: Debye mode first there is, then she disappears, then appears again, then vanishes again.

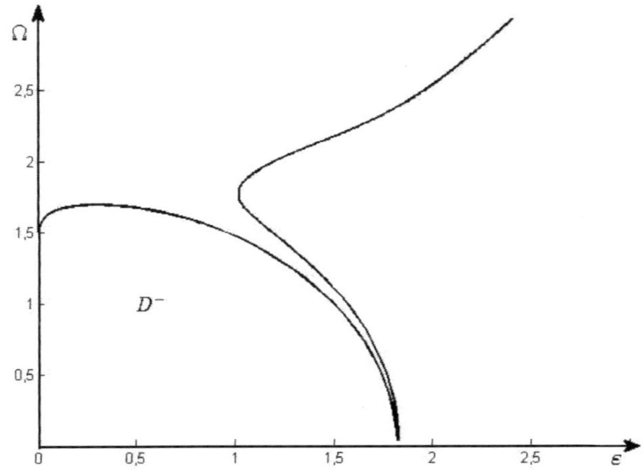

Figure 2. Curve L with α=30.

At about ε=1,8 radical expression for the curve L always vanishes, i.e. the graph will always touch the axis ε. When α=+∞ (fully degenerate plasma) the width of the loops on the graph will tend to zero. But it is a subject for further research.

3. Conclusion

Thus, it was shown that the Debye mode in the case of nearly degenerate electronic plasma shows a non-trivial dependence on parameters. These parameters include the frequency of collisions of electrons and frequency of the external electric field. This leads to some new features of screening of the external electric field by the plasma.

Region, where there is unusual behavior Debye mode, is very narrow. The width of this region decreases with the growth of the degree of degeneracy of electron plasma. For Maxwell plasma, this behavior is not observed.

These effects can manifest themselves in the interaction of electromagnetic waves with the surface of conductive medium or a conductive layer [7-10].

4. Acknowledgments

The work is performed under partial support of the Russian Foundation of Basic Researches and Government of Moscow Region (project No. 17-47-500752).

References

[1] Keller O. Local fields in the electrodynamics of mesoscopic media. *Physics Reports* 1996 **268** pp. 85-262.
[2] Girard C., Joachim C. and Gauthier S. The physics of the near-field. *Rep. Prog. Phys.* 2000 **63** pp. 893-938.
[3] Bhatnagar P.L., Gross E.M., Krook M. Model for collision processes in gases. I. Small amplitude processes in charged and neutral one component systems. *Phys. Rev.* 1954 **94** pp. 511-525.
[4] Landau L. On the vibrations of the electronic plasma. *Journal of Physics* 1946 **10** (1) pp. 25–34.

[5] Latyshev A.V. and Yushkanov A.A. Plasma in a high-frequency electric field with a reflective condition on the boundary. *Fluid Dynamics* 2006 **41** (1) pp. 161−172.

[6] Latyshev A.V. and Gordeeva N.M. The behavior of plasma with an arbitrary degree of degeneracy of electron gas in the conductive layer. *Theor. and Math. Phys.* 2017 **192** (3) pp.

[7] Kliewer K.L. and Fuchs R. Optical properties of an electron gas: further studies of nonlocal description *Phys. Rev.* 1969. **185** (3) pp. 905-913. 506–522.

[8] Jones W.E., Kliewer K.L. and Fuchs R. Nonlocal Theory of the Optical Properties of Thin Metallic Films. *Phys. Rev.* 1969 **178** (3) pp. 1201–1203.

[9] Yushkanov A.A. and Zverev N.V. Quantum electron plasma in one-dimensional metallic-dielectric photonic crystal. *Optics and Spectroscopy* 2017 **122** (2) pp. 202−206

[10] Yushkanov A.A. and Zverev N.V. Quantum Electron Plasma, Visible and Ultraviolet P-wave and Thin Metallic *Film Phys. Lett.* A 2017 **381** pp 679-684

IOP Conf. Series: Journal of Physics: Conf. Series **996** (2018) 012010 doi:10.1088/1742-6596/996/1/012010

The estimation of the internal dimensions of the channel for transport of atoms, molecules from the variation principles and the conditions of its local symmetry

N A Krylova, E V Kalashnikov

Moscow Region State University, Moscow 105005, Radio street,10, Russia

E-mail: krylova_nadya_1993@mail.ru

Abstract. Properties of channel in order to transport atoms (ions, molecules) is considered from point of its symmetry. Lagrange function of the transport channel is constructed. Properties of the Lagrange function and types of symmetry of the channel allow to estimate its inner size. Crystallographic channel of quartz and its dislocations are used as particular example.

1. Introduction

Variational principles, Dating back its roots to the works of Euler, Lagrange, Maupertuis [1] find wide application in modern physics, covering the research area from thermodynamics [2] and quantum physics [3,4].

One of the problems solved by L. Euler that is the problem (isoperimetric problem) search closed planar lines of finite length, which would restrict the maximum area [5].

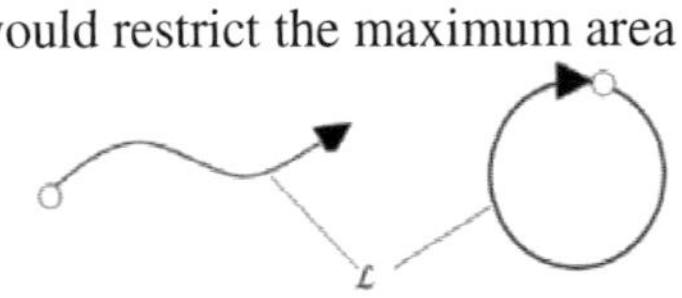

Figure 1. Flat line L is finite and fixed length bounds the maximum area on the flat surface

The scheme is "symmetry – shape - size" arises in problems related to the movement of the atoms (molecules, ions) through channels. For the migration of foreign atoms (ions, molecules) in various bodies play a great role transport channels, the transverse dimensions of which are comparable to or exceed times the size of the transported atoms (ions, molecules) [6 – 10]. Such channels can be crystallographic channels [10] (channels with open type crystallographic structure, for example, quartz, water), dislocation [6-10], tracks from charged particles, various molecular structures in biology [9]. The evaluation of the internal size of the channel (or the core of the dislocation) can be done [10,11], but the relationship of channel structure, its symmetry and its internal diameter remains unsolved. The search for such a connection will allow to determine the effect of the channel structure on the movement of the object within the channel.

Regardless of their functional properties, the channels exhibit some properties of symmetry. Among the most common types of symmetry of the channels can distinguish cylindrical and helical symmetry.

The aim of this work is to evaluate internal dimensions (radii) of the transport channel based on its symmetry properties.

2. The model and the solution

Consider the relationship of symmetry of the channel with its inner dimensions on a particular example of crystal channel, quartz (fig.1)

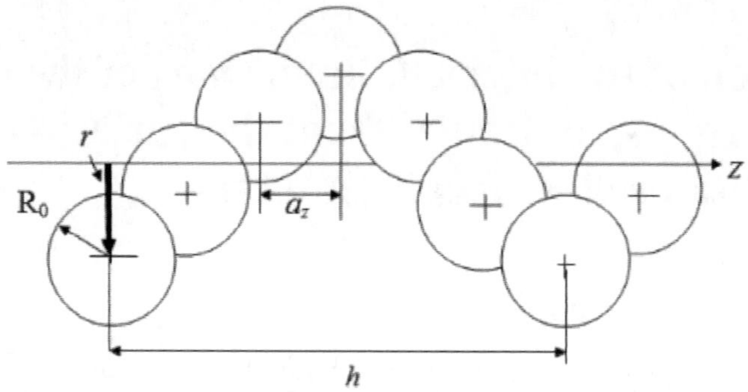

Figure 2. Crystallographic channel of the quartz along the direction [0001]. Depicted oxygen ions form the "inner wall" of the channel. Each ion belongs to the structural component of the quartz - tetrahedron SiO_4. R_0 is the radius of the oxygen ion; az is the projection on the z-axis distance between the centers of the oxygen ions in the channel; h - period; r is the distance from the axis of rotation to the center of gravity of the oxygen ion, forming a "wall" of the channel.

This channel exhibits properties of helical symmetry. To reveal these properties of symmetry, let us consider the Lagrangian $\mathcal{L}$ for a channel:

$$\mathcal{L} = \mathcal{L}(\vec{r}_1, \vec{r}_2, \dots, \vec{r}_n, \varphi_1, \varphi_2, \varphi_3, \dots, \varphi_n; \dot{\vec{r}}_1, \dot{\vec{r}}_2, \dot{\vec{r}}_3, \dots, \dot{\vec{r}}_n, \dot{\varphi}_1, \dot{\varphi}_2, \dot{\varphi}_3 \dots, \dot{\varphi}_n, t)$$

Here $\vec{r}_j$- position of j-th atom in the channel (Fig.1), φ_j- rotation of the j-th atom along the z-axis, the upper dot marks the derivative at time t from appropriate variables. The requirement of symmetry of a certain type implies the invariance of the Lagrange function relative to the variable for which the Lagrangian is preserved (Noether theorem).

At helical symmetry this means the variation of the Lagrangian $\delta\mathcal{L}$ of the channel with the translation along the z-axis (δz) and simultaneous rotation around the axis ($\delta\varphi$) equals to zero:

$$\delta\mathcal{L} = \sum_a \left(\partial\mathcal{L}/\partial\vec{r}_a\right)\delta\vec{r}_a + \sum_a \left(\partial\mathcal{L}/\partial\varphi_a\right)\delta\varphi_a = 0$$

(The cylindrical symmetry implies independence in relation to the transfer along the symmetry axis and the rotation about this axis

$$\delta\mathcal{L} = \sum_a \left(\partial\mathcal{L}/\partial\vec{r}_a\right)\delta\vec{r}_a = 0$$

and

$$\delta\mathcal{L} = \Sigma_a\left(\partial\mathcal{L}/\partial\varphi_a\right)\delta\varphi_a = 0\ )$$

Variation $\delta\vec{r}_a$ and $\delta\varphi_a$ for each *a*-th atom is the same, finite and arbitrary. Since each atom channel (Fig.1) is shifted by the same distance δz and rotated by the same angle $\delta\varphi$, then

$$\sum_a \left(\partial\mathcal{L}/\partial z_a\right)\delta\vec{z}_a = \delta z \sum_a \left(\partial\mathcal{L}/\partial z_a\right)$$

$$\sum_a \left(\partial\mathcal{L}/\partial z_a\right) = \left(\partial\mathcal{L}/\partial z_1\right) + \left(\partial\mathcal{L}/\partial z_2\right) + \dots + \left(\partial\mathcal{L}/\partial z_n\right)$$

Take into account that $z_n = z_1 + n \cdot a_z$ then $\partial z_n = \partial(z_1 + n \cdot a_z) = \partial z$ and

$$\sum_a \left(\partial\mathcal{L}/\partial z_a\right) = \left(\partial\mathcal{L}/\partial z_1\right) + \left(\partial\mathcal{L}/\partial z_1\right) + \dots + \left(\partial\mathcal{L}/\partial z_1\right) = n \cdot \left(\partial\mathcal{L}/\partial z\right)$$

The same with respect to variations $\delta\varphi$:

$$\sum_a \left(\partial\mathcal{L}/\partial\varphi_a\right) = \partial\mathcal{L}/\partial\varphi_1 + \partial\mathcal{L}/\partial\varphi_2 + \dots + \partial\mathcal{L}/\partial\varphi_n = n \cdot \partial\mathcal{L}/\partial\varphi$$

In this case.

$$\delta\mathcal{L} = n\left(\left(\partial\mathcal{L}/\partial z\right)\delta z + \left(\partial\mathcal{L}/\partial\varphi\right)\delta\varphi\right) = 0 \qquad (1)$$

With respect to (1) we note that the sign of the variation of a Lagrangian is not important, but the main role plays the minimum of the action $S = \int_{t_1}^{t_2} \mathcal{L}\, dt$ [12].

This means that the variation with positive or negative sign will inevitably lead to an increase in S. Therefore, the expression (1) is written in another form:

$$\delta\mathcal{L} = n\left(-\left(\partial\mathcal{L}/_{\partial z}\right)\delta z + \left(\partial\mathcal{L}/_{\partial\varphi}\right)\delta\varphi\right) = 0 \qquad (1a)$$

or

$$\delta\mathcal{L} = -(\dot{P}_z)\cdot\delta z + (\dot{L}_z)\cdot\delta\varphi = 0$$

Taking into account

$$\delta z = \left(h/_{2\pi}\right)\delta\varphi$$

one has

$$\delta\mathcal{L} = \frac{d}{dt}\left(-P_z\cdot\left(h/_{2\pi}\right) + L_z\right)\cdot\delta\varphi = 0$$

then the value

$$-P_z\cdot\left(h/_{2\pi}\right) + L_z = 0 \qquad (2)$$

is a conserved quantity. The constant of integration is chosen equal to zero. This eliminates uncertainty L_z, depending on the size of sought value r .

$$L_z = mr^2\dot{\varphi} = mr^2\omega \qquad (3)$$

Note that if the variation was preserved in the form of (1), (2) would receive the complete positive form, It would lead to non-physical (imaginary) values of the radius of the channel (4), which is not good for us. Therefore, substituting (3) into (2), we obtain

$$r^2 = \left(mv_z\cdot h/_{2\pi m\omega}\right) = v_z\cdot h/_{2\pi\omega} \qquad (4)$$

r means the distance from the axis of rotation of the channel to the center of gravity of the oxygen ion, forming a "wall" of the channel (Fig.2). It can be written

$$r^2 \approx \left(\Delta z/_{\tau}\right)\cdot h/_{2\pi\left(2\pi/_{\tau}\right)} = \Delta z\cdot h/_{(2\pi)^2}$$

Here Δz – displacement along the axis z, τ – the time during which the channel would be rotated and shifted to the step h. Taking $\Delta z = h$, we get

$$r = h/_{2\pi} \qquad (4a)$$

Here it is possible to transit to the "language" of burgers vectors of $b(h = 2b)$. In the model of the channel, *Fig.2*, $h = 6a_z$ and

$$r = b/\pi \qquad (5)$$

Let estimate the radius for a channel with cylindrical symmetry. It is impossible to estimate radius directly from (1) under the condition (3). (Loop should be closed). To estimate the radius of channel with cylindrical symmetry, we assume that the spiral is curved in a single step a_zi.e., $h = a_z$.

In this case, the minimum value of the ratio of the radius of channel with helical symmetry to the radius of the cylindrical symmetry equals

$$\left(6a_z/_{2\pi}\right)\Big/\left(a_z/_{2\pi}\right) = 6 \qquad (6)$$

The value of the radius (of the centers of gravity of the ions forming the inner wall of the channel) for a channel with helical symmetry (Fig.2) equals to

$$r = 6a_z/2\pi = 6\cdot 1{,}83\dot{A}/6{,}28 \approx 1{,}75\dot{A} \qquad (7)$$

The radius of the cavity channel, $r - R_o$, (Fig.2) can be positive or negative. Here, the ion radius of the oxygen atom $R_o = (1{,}2 \div 1{,}3)\dot{A}$ [13]. Then the "radius" of the cavity of the channel for screw symmetry is equal to:

$$r_z = 1{,}75\dot{A} - (1{,}2 \div 1{,}3)\dot{A} = (0{,}55 \div 0{,}45)\dot{A} \qquad (8)$$

and for cylindrical symmetry we get the following value

$$r_c = a_z/_{2\pi} - (1{,}2 \div 1{,}3)\dot{A} = 0{,}291\dot{A} - (1{,}2 \div 1{,}3)\dot{A} \approx -(0{,}9 \div 1{,}0)\dot{A} \qquad (9)$$

From comparison of relations (8) and (9) of the radii of the channels for screw and cylindrical symmetry it follows that the channel with cylindrical symmetry is covered with oxygen ions is much stronger than the channel with helical symmetry, as can be seen in figure 3.

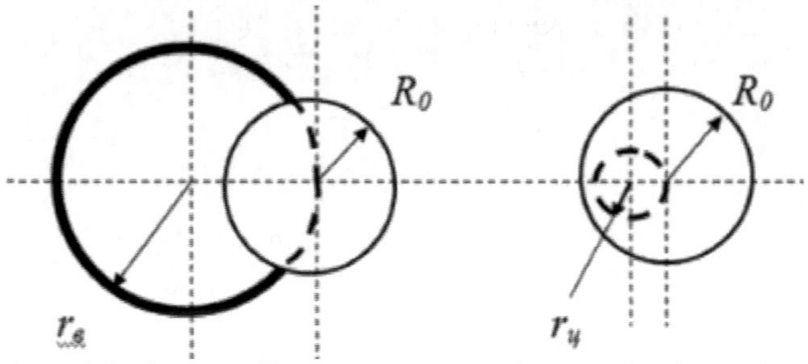

Figure 3. The overlap of the ionic radii of oxygen crystallographic channel: a) helical symmetry and b) with cylindrical symmetry. The dotted line indicates the blocked part of the channel. R_0 is the radius of the oxygen ion; rв is the radius of channel with helical symmetry; rц is the radius of channel with cylindrical symmetry.

This implies that in a channel with helical symmetry "free space" for the placement of foreign atom more than in a channel with cylindrical symmetry. Minimum screw dislocation (providing the minimum elastic energy) in quartz are located along the crystallographic channels by stretching or compressing them [10]. Distance a_z (7) and (9) along the z-axis of "expanded" channel is changed :$a_z \Rightarrow (3/2)a_z$. The ratio (6) will not change, but the radii of the cavities (8) and (9) for both helical and cylindrical symmetry will change as follows:

1) for the core of the dislocation with screw symmetry the radius of the distance of the centers of gravity of the oxygen ions from the z axis is equal to $r = 2{,}65\mathring{A}$ the radius of the cavity

$$r_z = (3/2) \cdot 1{,}75\mathring{A} - (1{,}2 \div 1{,}3)\mathring{A} = (1{,}425 \div 1{,}325)\mathring{A}$$

2) for the same channel in the approximation of cylindrical symmetry associated radii equal to: $r = 0{,}291\mathring{A}$

$$r_ц = (3/2) \cdot a_z/2\pi - (1{,}2 \div 1{,}3)\mathring{A} = 0{,}291\mathring{A} - (1{,}2 \div 1{,}3)\mathring{A} = -(0{,}763 \div 0{,}863)\mathring{A}$$

Grown nanotubes [15] have demonstrated properties of the cylindrical and helical symmetry too and have different radii

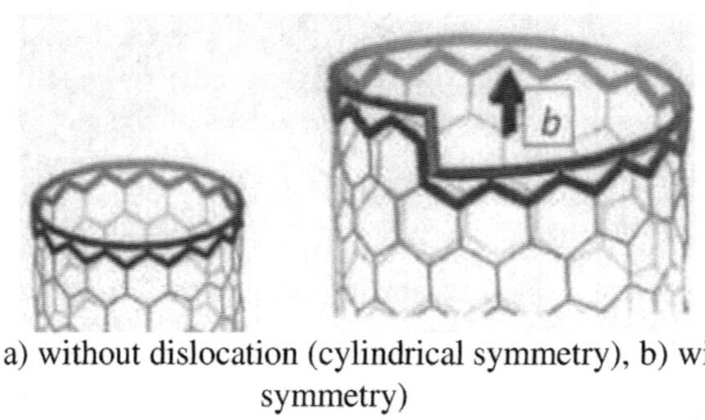

Figure 4. Nanotubes a) without dislocation (cylindrical symmetry), b) with dislocation (helical symmetry)

3. Conclusion

The work was able to show, explicitly, the influence of the type of the local symmetry of the transport channel at its inner radius of the cavity. It turned out that the channel with helical symmetry has an internal radius substantially greater than the radius of the channel with cylindrical symmetry. Since the clarification of connection of local symmetry with form and size of channel was used General principles of minimum of action, it can be assumed that, in most cases, the organization of transport channels for the transfer of atoms (molecules, ions) preferred type of symmetry is the structure of the channel showing the properties of helical symmetry.

References

[1] Variational principles of mechanics. Collection of articles: Hamilton, Euler, Gauss, etc / L. Polak. (ed.). — M.: Fizmatlit, 1959. 932 p.
[2] Gyarmati: Non-equilibrium Thermodynamics. Field Theory and Variational Principles. Springer-Verlag, Berlin, Heidelberg, NewYork 1970..
[3] Y. Shwinger, Particles, Sources, and Fields, v1. N.Y, USA: Prinst Publ. University, 1998.
[4] Y. Shwinger, Particles, Sources, and Fields, v2. N.Y, USA: Prinst Publ. University, 1998.
[5] I.M. Gelfand, Fomin C.V. Calculus of variations. M.:Izd.Fizico-matematicheskoj Literaturi, 1961, 228 p.
[6] P.Sui , Z. Dai . "Nonlinear dynamical behavior of Xenon atoms along dislocation lines in UO_(2+x) nuclear fuel." Science China Physics, Mechanics & Astronomy. 2015. **58**, 5, p.1.
[7] A.Ishii, J. Li, S.Ogata. "Conjugate Channeling" Effect in Dislocation Core Diffusion: Carbon Transport in Dislocated BCC Iron." Journ.Pone.0060586.PDF. 2013. **8**, 4, p.1.
[8] Wang, L. et al. Molecular valves for controlling gas phase transport made from discrete ångström-sized pores in grapheme // Nature Nanotechnol. 2015. **10**, p.785.
[9] L.Wang, et al. "Molecular valves for controlling gas phase transport made from discrete ångström-sized pores in grapheme" Nature Nanotechnol. 2015. **10**. p.785.
[10] J.-F. Gwan, A. Baumgaertner. "Ion Transport in a Nanochannel" Journal of Computational and Theoretical Nanoscience. 2007. **4**, p.1,
[11] E.V. Kalashnikov, I.N. Tolstikhin, B.Z.Pevzner. "Motion of Helium atom through quartz crystal with dislocations"Physics of the Solid State, 2010. **52**, 7, p.1283
[12] J.Friedel, Dislocations. Pergamon Press, Oxford, 1964, 643 p.
[13] L.D. Landau, E.M. Lifshitz. Mechanics, v.1. M.: Nauka, 1988, 215p.
[14] L. Pauling. Nature of the Chemical Bond and Structure of Molecules and Crystals. Second ed. Cornell University Press, Ithaca, NY, 1940. 429 p.
[15] Feng Ding, Avetic R. Harutyunyan, Boris Yakobson. "Dislocation theory of chirality-controlled nanotube growth". PNAS 2009. **106**, 8, p.2506.

Sources and components of ball lightning theory

A I Nikitin[1], V L Bychkov[2], T F Nikitina[1], A M Velichko[1], and V I Abakumov[2]

[1]Talrose Institute for Energy Problems of Chemical Physics, Russian Academy of Sciences, Moscow, Russia.
[2]Lomonosov State University, Moscow, Russia

E-mail: anikitin@chph.ras.ru

Abstract. The article describes the cases when ball lightning (BL) exhibited an extremely high specific energy store (up to 10^{10} J/m3), a presence of uncompensated electric charge (up to 10^{-3} C) and an ability to generate high frequency pulses (up to 10 MW). It is shown that the realization of a combination of these properties of BL is possible if to consider it as a heterogeneous system consisting of a unipolarly charged core and a dielectric shell. In the electric field of the core charge, arises a force owing to the polarization of the shell that opposes the Coulomb repulsion force of the charges. BL models constructed according to the indicated principle are described: the electrodynamic model and the chemical-thermal model, which treats BL as a hollow sphere filled with steam. The requirement to take into account the main three properties of BL makes it possible to reduce the number of models of this natural phenomenon. Detailed cases of observations of high-energy lightning are analyzed.

1. Introduction

Scientific investigation of BL problem has started in 1838 by publication of F. Arago's article [1]. Unfortunately, this investigation still did not led to understanding of physical nature of this natural phenomenon. A reason of this, apparently, is ignoring of key properties of BL and attempts to build its models on the base of some myths. These "myths" are: 1) an aspiration to build a model of BL in a form of homogeneous system stabilized only by internal forces and atmospheric pressure; 2) the statement that the average density of BL is close to air density; 3) the assumption of BL electric neutrality; 4) an attempt to explain BL luminescence only by thermal radiation of hot bodies [2-4]. It is time to define new basic properties of a BL which can form a basis of its theory creation.

In our opinion, ignoring of three major properties of BL –high energy storage, a presence of non-compensated charge and its ability to generate a large power radio emission – was a principal cause of failures of creation of BL theory. We consider the listed properties as key ones and we define on their basis the three base principles of its model developing – a heterogeneous structure, presence of charge carriers and ability of self-organizing of substance of BL core.

2. Cases of ball lightning energy measurement

The most mysterious property of BL is its ability to accumulate a big amount of energy in the limited volume. As the first documentary case of high energy BL observation one may consider the publication in the newspaper "The Daily Mail (London)" on November, 5th, 1936 of a letter to the editor named "A thunderstorm mystery" [5]. " Sir, during a thunderstorm I saw a large, red hot ball come down from the sky. It struck our house, cut the telephone wires, burnt the window frame, and

IOP Conf. Series: Journal of Physics: Conf. Series **996** (2018) 012011 doi:10.1088/1742-6596/996/1/012011

then buried itself in a tub of water which was underneath. The water boiled for some minutes afterwards, but when it was cool enough for me to search I could not find nothing in it". Dorstone, Hereford. W. Morris. The description of a case, published in the newspaper, became known to scientific community. This was during a discussion at the Institution of Electrical Engineers in London on January, 7th, 1937, following the presentation of a review paper by Professor B.J. Goodlet, entitled "Lightning" [6]. During the discussion Professor Goodlet estimated the energy content of the ball. He estimated that Mr. Morris's description of the heating of the water suggested a temperature rise to 60^0 C. Thus, if the initial temperature was 20^0 C, $\Delta\theta = 40^0$ C. Four English gallons of water have a mass m_1 = 18 kg. The specific heat capacity of water $c = 4200$ J·kg^{-1}·K^{-1}, therefore $\Delta Q = m_1 c\Delta\theta = 3.1$ MJ. After that he carried out other calculation based on the assumption, that all the water has heated up to 100^0 C, and a mass $m_2 = 1.8$ kg was evaporated. The latent heat of evaporation of water $L = 2.26$ MJ/kg, hence, $\Delta Q = m_1 c\Delta\theta + m_2 L = 10$ MJ. If to accept diameter of the ball equal to 10 cm, and energy equal to 3.1 MJ, the energy density of the ball can be estimated as $0.6\cdot10^{10}$ J/m^3. For the second variant the density of energy of ball lightning will appear equal to $\rho_E = 1.9\cdot10^{10}$ J/m^3.

The "natural experiment", described by Mr. Morris, was not single. In August, 1962 in Ukraine near the city Perechin BL with the size of tennis ball fell in a trough with water for cattle. The water was almost completely boiled away from the trough, the cooked frogs laid at the bottom. The size of the trough was 0.3×2.5 m, depth of water layer was 15 cm [7, pp 106]. Accepting that mass of water in the trough was $m_1 = 112.5$ kg, mass of the evaporated water was $m_2 = 100$ kg, and water heating occurred from 10°C to 100° C, we find, that energy, transferred to water by BL, was equal to 269 MJ. The volume of a tennis ball 6 cm in diameter is equal to 113 cm^3, from here the energy density of the "Ukrainian" BL was $\rho_U = 2.38\cdot10^{12}$ J/m^3. This density is approximately in 100 times larger than energy density of the "English" BL. The error in definition by the observer of the size of the ball could become the reason of it. If its diameter was equal to 28 cm, the value of energy density would be nearby $2.4\cdot10^{10}$ J/m^3.

Next important event occurred in July, 1972 in Hungary [8]. At midday in presence of numerous factory workers a brightly shining ball the size of soccer ball fell in a pit with 120 liters of water. All the water has evaporated. For heating of 120 liters of water by 80°C and its evaporation the energy 311 MJ is necessary. Accepting the diameter of the ball equal to 25 cm, we find the energy density of the "Hungarian" BL $\rho_H = 3.8\cdot10^{10}$ J/m^3. Thus, it is necessary to regard the cases, when the energy of BL has been measured with the help of "water calorimeter", as the reliable facts. Below we describe two new cases of estimation of BL energy.

Here is a story of engineer N.F. Ignatov, the participant of event that has occurred in July, 1964 at the river Ucha near the stop Mamontovka in the Pushkin district of the Moscow Region. The story-teller was then 20 years old [9]. "At that time I was the student of technical school of communication. At 4 o'clock in the morning I was fishing at the river U_{cha}. The sun has not risen yet, the weather was clear, there was no rain this morning and overnight. In the place of fishing the width of the river was 20-30 meters. I had a simple reed fishing-rod with an ordinary line and hook. Before the event I have caught 5-6 roaches which I held in the 10 l Gi pail. The pail was half-filled with water. I felt that a fish had eaten a worm from a hook. I took out a line with an empty hook from water and when I was carrying a hook, hanging on the line, above the water's edge (the hook was about 1 meter above the ground), a shining ball with diameter about of 5 centimeters suddenly "clung" to it. I did not notice from where the ball had arrived. The weight of a sinker with the hook when the ball had stuck to it did not increase. I don't remember the color of the ball, it was shining slightly brighter than a luminescent lamp, and its light did not blind eyes. The ball contour was accurate. I was confused and slowly lowered the hook with the ball, hanging on it, into the pail with the caught fish. At once the water began to hiss and gurgle, as it happens, when the heated stone is thrown into it. Water was boiling for some more minutes. Fishes were cooked so as it had been boiled too long: it had been skinned, scaled and gutted. At boiling a dense steam rose from water. The hook, tied to the line, had disappeared. The line remained untouched, it became only slightly shorter. In fear from uncertainty of what had occurred, I upset the remains of the water with fish on the ground and came to a hostel with the empty pail. When I told my companions about the incident, they did not believe me: they decided that I had caught nothing and I was justifying myself for it. So I had to go and show them the boiled fish on the ground."

Let's estimate energy which was brought by BL in the pail with the caught fish. The capacity of a standard pail, made of Gi, we will accept equal to 10 liters. There was in it $m_w = 5$ kg of water and $m_f = 0.9$ kg of roaches (6 fishes with an average weight 150 g). The initial temperature of water we will accept equal to 20° C. Let the specific heat capacity of the fish is equal to specific heat capacity of water $c = 4.2$ kJ/kg·K. The energy, spent for water heating by $\Delta\theta = 80^0$ C before boiling, is $Q_w = c(m_w + m_f)\Delta\theta = 1.98$ MJ. It is necessary to add to this result the heat Q_p, spent for heating of the pail with mass $m_p = 1$ kg by $\Delta\theta = 80^0$ C. A specific heat capacity of steel is $c_s = 0.46$ kJ/kg·K, therefore $Q_p = c_s m_p \Delta\theta = 36.8$ kJ. As a result we obtain that energy of the "Moscow" BL appeared equal to $Q_M = Q_w + Q_p = 2$ MJ. It is the lower estimate of energy. The eyewitness spoke about a production of steam at water boiling, on what, naturally, the part of energy of BL was spent. But, because data about quantity of the evaporated water is not available, we can tell nothing about amount of this additional energy. The volume of a ball of diameter 5 cm is equal to 65 cm^3, from here we find, that the energy density in the ball lightning was not less than $\rho_M = 3\cdot10^{10}$ J/m^3. As we see, this figure is well coordinated with the values of energy density of "English" ($\rho_E = 1.9\cdot10^{10}$ J/m^3) and "Hungarian" ($\rho_H = 3.8\cdot10^{10}$ J/m^3) BL. Apparently, the fishing hook has disappeared because of action on it of the high-frequency electromagnetic field, radiated by BL. The observers repeatedly informed about disappearance of metal subjects (rings, bracelets, etc.) after their meeting with BL [7, 8, 10, 11]. It is probably that water heating also occurred because of action of this electromagnetic radiation on it.

On May 27th 2013 at 15 o'clock of local time BL destroyed a house in the outskirts of village Mogsokhon of Kizhinginsky district of Buryatia [12]. According to stories of eyewitnesses of the event, "A usual rain began, but suddenly a thunder struck with such a force that passers-by and the cattle, being at this time in the street, crouched with horror and fled their several ways. A minute later after that the bright shining ball descended from the sky and forced its way into the house of the Bayan Sandanov's family through a roof. Then there was a deafening explosion inside the house. A housewife, the spouse of Bayan, cleaning the kitchen after dinner, was in the house during the explosion. Found under pieces of a fallen wall she was seriously bruised and lost partially her hearing. The suffered woman was hospitalized. After the explosion the house was in a critical condition: doors and windows were smashed and thrown away at a large distance, the wall part fell, and the roof was partially destroyed (see Fig. 1). Because of BL explosion in many neighbor houses electric appliances were destroyed".

Let's try to estimate the energy of BL, which exploded in the house. We carry out estimation on a basis of definition of trinitrotoluene mass, which explosion leads to consequences similar to that, which occurred in the village Mogsokhon. According to a demolition guide [13, pp 125], for destruction of a flat package of logs a weight C (in grams) of a contact charge (directly located on the object to be undermined) is defined by a formula:

$$C = kF, \tag{1}$$

where F is an area of cross-section of a package (cm^2), and a factor k is defined by a kind and a condition of wood. For a dry pine or a spruce $k = 1$. Let us accept the height of the wall, which fell out of the house, equal to $H = 3$ m. This wall is constructed out of 13 logs, from here we find a thickness of a log $d = 23$ cm and the cross-section of package $F = Hd = 6900$ cm^2. From the formula (1) at $k = 1$ we find $C = 6.9$ kg of trotyl. A heat of trotyl explosion is equal to 4 MJ/kg [14, pp 250], so the energy, containing in 6.9 kg trotyl, is equal to 28 MJ. If blasting is carried out by a charge which is at a distance r (m) from a wooden wall, the weight of the charge C(kg) can be found from a formula [13, pp 127]:

$$C = 30kdr^2. \tag{2}$$

Here d (m) is a thickness of a log. If the charge is in the centre of a room of size 4×4 m^2, then at $r = 2.2$ m and $d = 0.23$ m we find $C = 33$ kg trinitrotoluene, that is equivalent to energy of 132 MJ.

In the description of the event nothing is said about a size of BL. If to accept it equal to a diameter of “average” BL D = 30 cm, then the energy density of the “Buryat” BL ρ_B will be from $2.1 \cdot 10^9$ J/m^3 to $9.3 \cdot 10^9$ J/m^3. These figures are of the same order, as values of energy density of highly energetic BL, about which it was spoken above. All that allows to say that BL is dangerous natural phenomenon. This conclusion is not coordinated with opinion of some researchers, who speak about insignificant danger of BL. For example, Stenhoff [5, pp 94] considers it as a non-dangerous object and all destructions, ascribed to it, explains by action of usual linear lightning.

Figure 1. The house wall thrown out by the explosion

3. Cases of pulse emitting of ball lightning energy

In [7] a case of golden bracelet evaporation by BL is described: “There was a thunder-storm. A woman was sleeping in a house, which doors and windows were reliably locked inside. She was woken by a sound of a lightning stroke similar to a shot. Coming to a window, she with a surprise has found a small hole in it with a size of fifty-kopeck piece (5 cm). And suddenly she had noticed that the golden bracelet has disappeared from her hand. On a wrist there was only a dark strip. (The analysis has shown that it was a gold oxide). Very careful searches gave nothing. Only a round fragment of glass with melted edges was found on a floor. It precisely coincided with the hole in window glass. A dog, killed by lightning, was found in a kennel near the house”. Let us estimate an energy spent by BL on evaporation of the bracelet. We will accept a mass of the bracelet m_{br} equal to 50 g. The energy Q_1 of the bracelet heating from the room temperature $T_0 = 20°$ C to the temperature of gold melting T_m = 1063.4° C is equal to $Q_1 = C_{Au} \cdot m_{br} \cdot (T_m - T_0) = 6.78 \cdot 10^3$ J. (Here C_{Au} = 0.13 J / (g·K) is specific thermal capacity of gold). Energy Q_2 of the bracelet melting is $Q_2 = m_{br} \cdot l_{Au} = 3.2 \cdot 10^3$ J. (l_{Au} = 63.97 J/g is the latent heat of gold fusion). For increasing of the melt temperature to the boiling point $T_b = 2887°$ C the energy Q_3 is necessary, $Q_3 = C_{Au} \cdot m_{br} \cdot (T_b - T_m) = 11.8 \cdot 10^3$ J. Specific heat of gold evaporation is L_{Au} = $1.68 \cdot 10^3$ J/g, from here the bracelet evaporation energy is $Q_4 = m_{br} \cdot L_{Au} = 84 \cdot 10^3$ J. Total energy of the bracelet evaporation is $Q = Q_1 + Q_2 + Q_3 + Q_4 = 10^5$ J, and for evaporation of one gram of gold the energy $2.11 \cdot 10^3$ J is necessary. If to accept the time of evaporation of the bracelet equal to 10^{-2} s, then a power of the energy source should be no less than P = 10 MW.

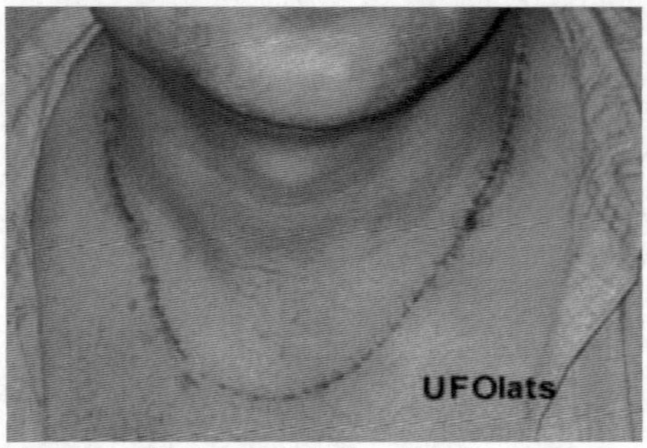

Figure 2. The trace of burn, made by the evaporated golden chain.

On July10, 2006 in Jürmala city in Lithuania when a resident of Yelgava city Oleg Andrejev went from the sea to the coast, "a yellow-orange bright ball about of a tennis ball in size has flown" nearby him and "a strong roar has appeared. Oleg has fallen face downwards the sand. He did not lose consciousness, but he was very sluggish, its consciousness became disseminated. When Oleg was turned, his face has appeared covered with black soot. People, bathing at this time in the gulf, had felt appreciable current blow ... The golden chain on Oleg's neck had been evaporated, having left the appreciable burn round his neck in the form of a dotted line. There were traces of action of the lightning on his stomach and his heels" [11, pp 234] (see Fig. 2). If the mass of the chain was 17 g, the energy of its evaporation was nearby 36 kJ, and the power (at the explosion duration 10^{-2} s) was 3.6 MW.

In [15] a case of "theft" of ring metal subjects by BL is described: "In a barn on hay two – a man and a woman were sitting. Suddenly a body, more large than an apple, directed to a whiteness, in zigzag fashion. It had swum to them for a distance of one meter. Its body was sparkling, and the woman's hairs (their length was 20-30 cm) had risen on end (as a fan), and the man's hairs begun to move under a peak-cap ... The ball had left the barn through a crack in a roof ... Recovered, the observers had found, that the woman's rings on fingers had disappeared, and the man's chain of a penknife, a metal tip of a pen and metal rims on the boots, fixing edges of apertures for laces," also had disappeared. If a reason of the specified subjects disappearance was their evaporation by BL energy, then for evaporation of a golden ring with mass 14 g the energy 30 kJ is necessary, and for the all gone subjects –it is several times more (nearby 100 kJ). Notice that the BL had spent for performing of the specified action only a part of its energy.

On August 12 1978 at 23 o'clock 20 minutes in Khabarovsk near Hasanovsky street during a pouring rain the spectators went out of the cinema "Zarja", had heard a sharp whistle, reminding a work of jet engine. The whistle was accompanied by crash. It became light, as in the afternoon. Then over a cinema building BL of about 1.5 m diameter had appeared. A color of BL was orange, from it sparks were strewed. The BL had started to descend, had gone down to the earth surface through branches of trees, for an instant it had flashed over the earth and again had risen upwards. A strong explosion was heard, it became dark and silent. BL was observed during about one minute. On a distance up to 100 m electric wiring had been put out of action. Despite the torrential rain and wet soil, the soil's site of 1.5 m diameter and depth of 20-25 cm had been charred, and a slag had been formed in it. The volume of the zone, filled with the slag, was nearby 0.4 m^3. The slag consisted not of solid crust, but of many pieces of irregular form with the average size of 5-6 cm. Some pieces were alloyed to each other. In total it had been found about a thousand of such pieces. The initial ground represented a sandy loam on the basis of quartz sand [16].

Authors of [16] had made experiments for definition of a method of power action on the soil, which result was a formation of the slag similar to the slag appeared at BL explosion. They found that similar results of action on the soil had been obtained at its heating by high-frequency radiation at frequency 13.56 MHz and power 60 kW. Authors had estimated the energy released at explosion of BL. BL action had led to heating to fusion temperature (T_m = 1700° C) m_g = 440 kg of the ground in a volume of the cylinder of diameter 1.5 m and height of 0.22 m and had evaporated m_w = 175 kg of water from this volume. Having accepted a specific thermal capacity of the ground equal to $C_g = 0.84$ kJ/(kg·K), we find the energy, spent for its heating from the initial temperature $T_0 = 20°$ C to the temperature of fusion: $E_g = C_g \cdot m_g \cdot (T_m - T_0) = 6.2 \cdot 10^8$ J. For heating $m_w = 175$ kg of water from 20° C to 100° C energy $E_{hw} = C_w \cdot m_w \cdot (100^0 - 20^0) = 0.59 \cdot 10^8$ J had been spent. ($C_w = 4.2$ кJ/kg·K is specific thermal capacity of water). Specific heat of water evaporation is $L_w = 2.26 \cdot 10^6$ J/kg [17 (35)]. For evaporation of 175 kg of water an energy $E_{vw} = 3.95 \cdot 10^8$ J is needed. Thus, at a flash of BL with the diameter of 1.5 m the energy $E_t = E_g + E_{hw} + E_{vw} = 10.7 \cdot 10^8$ J had been released. The BL volume was $V_{bl} = 1.77$ м^3, from here a density of its energy is $\rho_E = E_t/V_{bl} = 6 \cdot 10^8$ J/m^3. If to accept a duration of flash of BL over the earth equal to 0.1 s, then the power of electromagnetic radiation will be $P = 10^{10}$ W. This very day in other area of Khabarovsk a ball lightning has flown in a tank with $m_{w2} = 7$ t of water and has heated it up to boiling [18 (36)]. Proceeding from it, its energy was $E_{bl2} = C_w \cdot m_{w2} \cdot (100° - 20°) = 2.35 \cdot 10^9$ J.

The cases described above allow to draw a conclusion on BL ability to generate high-frequency radiation. The energy of the radiation pulse in some cases exceeds 10^5 J that is 1-10 % of the average sized (10-20 cm) BL energy. Some representation about of wave lengths range of BL electromagnetic radiation can give a case of its observation by spouses Popele on January 6 2011 near Budapest in Hungary [19, pp 300]. Spouses managed to photograph BL laying on snow (see Fig. 3).

Figure 3. Photo of BL on snow.

The diameter of BL was nearby 50 cm. At the left and to the right of BL between it and the snow air luminescence has been observed. It is not excluded, that it had been caused by a corona discharge. The most surprising in described event was that the spouses, investigating after disappearance of the BL the place where it was laying before, had not found any traces of snow thawing. The reason of this can be that BL radiation lays in a range of radio-waves not absorbed by snow. For check of this idea we have placed a polyethylene box with snow into a microwave stove. At action on snow during 30 seconds of radio emission with the power of 850 W it remained untouched. This means, that the range of BL radiation wave lengths can be in the region 1-13 cm (the radiation wave length at which microwave stoves work, is equal to 12.25 cm) [19, pp 300].

4. Cases of manifestation of ball lightning electric charge action

In the previous section we told how BL having come at the distance of one meter to a woman's head, had forced like a "fan" to rise her hairs. In [20] a similar case, told by E.N. Kudrjavtseva, is described: "In the summer of 1947 I was on tours with an ensemble in the Chernovtsy city in the Moldavian Soviet Socialist Republic. After rehearsal at 2-3 o'clock p.m., I came to a colleague to the house where she was temporarily living. It was a small one-storey house. We sat by a window opposite each other and were knitting. The window leaf was opened. The thunder storm in the street gathered, lightnings sparkled somewhere away, but here still there was calm. Suddenly I have looked at the girlfriend and have seen that her hairs have risen upwards (on end). She was looking at me with the eyes expanded from a fright because with me, as it has appeared, there was the same. We then have told it to each other, and everyone on oneself did not feel anything unusual. Here we have seen that over us a ball of the diameter of 8-14 centimeters was floated. It was matte as a matte bulb, it was shining with brightness of the full moon. It moved very slowly. The room was small: the furnace was opposite the window, and in the room's centre a bulb (40 W) was hanging on a long cord. When the ball was flowing near the lamp, the incandescent hair was lighted to reddish color though the switch was switched off (we have checked this). When the ball has flown away from the lamp, the incandescent thread has gone out also. Then the ball has flown near the oven, cautiously bypassing all corners and not touching it, has turned along the wall and again began to come nearer to us. We were sitting, not moving, not talking. When the ball has flown away from us the first time, our hairs have fallen down. Now at its approach they have risen again on end. The ball has floated over our heads and has taken off for a window leaf. Soon after that the thunder storm has burst, the rain poured, as from a bucket, but all has quickly ended". If to accept a distance between a floor and a room ceiling equal to 2.3 m and to consider, that BL moved horizontally at a level of a window leaf height (nearby 2 m), it has passed at a distance of 1 m from the girls' heads.

At popular lectures about electricity action an effective experiment is often shown. The girl with the dismissed hairs stands on an isolating support and she is suggested to touch a pole of the electrostatic generator by a conducting core. A wire, lying on a floor, is attached to the other pole. At increasing of a voltage up to 300 kV the girl's hairs rise in the form of a fan. Let us estimate, at what value of an electric field intensity near to the girl's head there is a force, capable to rise her hairs. Let us replace the girl's figure with an ellipsoid of revolution with a size of big axis $2a$ = 1.6 m and a size of a small axis $2b$ = 0.3 m. An electric potential on the big axis at a distance x from the centre of the conducting ellipsoid is defined by a formula [21]:

$$\varphi = \frac{q}{4\pi\varepsilon_0 (a^2 - b^2)^{1/2}} Arth\left(\frac{a^2 - b^2}{x^2 + a^2}\right)^{1/2}, \tag{3}$$

and its capacity is:

$$C = [4\pi a_0 (a^2 - b^2)^{1/2}] \,/\, Arch\,(a/b). \tag{4}$$

Here q is the charge of the ellipsoid, and $\varepsilon_0 = 8.85\cdot10^{-12}$ F/m is the electric constant. Electric field intensity E on the ellipsoid surface (at $x = a$), directed along the big axis of the ellipsoid, is:

$$E = -\,d\varphi/dx = q \,/\, [4\pi\varepsilon_0\cdot\sqrt{2}(a^2+b^2)]. \tag{5}$$

At a = 0.8 m and b = 0.15 m the capacity $C = 4\pi\varepsilon_0\cdot0.333$ F. At the ellipsoid potential $U = 3\cdot10^5$ V its charge is $q = C/U = 4\pi\varepsilon_0\cdot1.0\cdot10^5$C. From the formula (5) we find the intensity of electric field at the "top" of the ellipsoid $E = 10^5$ V/m. Thus, for hairs raising the electric field intensity on the surface of the girl's head should be no less than 1 kV/cm. This conclusion was checked up experimentally. For this purpose in the center of the electric condenser formed by two horizontal metal disks with diameters of 105 and 160 mm, displaced from each other at 50 mm, a piece of a cotton thread of 0.3 mm diameter and 45 mm length was placed. The thread was attached by one end to the bottom electrode. At voltage 10 kV the thread took a vertical position. Thus, for its raising electric field E_{up} = 10 kV/5 cm = 2 kV/cm is needed [22].

Now, knowing the value E_{up}, we define a charge Q of BL for two described above cases. In both cases BL with the radius R_{bl} = 4 cm has created the electric field intensity $E_{up} = (1\text{-}2)\cdot10^5$ V/m at the distance R = 1 m. BL was indoors, therefore with good approximation it is possible to consider that the configuration of its electric field was close to a configuration of the spherical condenser field. For such a condenser the electric field intensity depends on the distance R to the charge centre as $E = Q/(4\pi\varepsilon_0 R^2)$ [23]. Assuming $E = E_{up} = (1\text{-}2)\cdot10^5$ V/m and R = 1 m, we find $Q = 4\pi\varepsilon_0 R^2 E_{up} = 10^{-5} - 2\cdot10^{-5}$ C. For such a charge the electric field intensity on the surface of BL (at $R_{bl} = 410^{-2}$ m) is equal to $E_{bl} = 6.210^7$ V/m – $1.2\cdot10^8$ V/m. It is by 20-40 times greater than the field intensity of air breakdown $E_{br} = 3\cdot10^6$ V/m. Despite this, in both cases there was not information about a discharge presence between

BL and nearby subjects even when it was passing at a short distance from walls and metal subjects like a damper or furnace rings.

The reason of confinement of large charge of BL is, apparently, a presence of a good insulator in area between the charge carrier and its surface [22]. However, at shorting of this insulator with any conductor the current of BL discharge can cause fusion of a conductor or an arc discharge [5, 24, 20]. If the discharge current passes through a body of a person, it can lead to mutilations and even to death [24, 10, 20, 19, 25]. Let's define a risk to which the girls, observing BL with charge $Q = 2\cdot10^{-5}$ C were exposed. The electric capacity of a ball of radius $R_{bl} = 4\cdot10^{-2}$ m is equal to $C_{bl} = 4\pi\varepsilon_0 R_{bl} = 4\cdot10^{-12}$ F. Accepting a resistance of the person's skin R_s equal to 1 kOhm, we find a characteristic time of pulsed discharge of the ball $\tau_{bl} = R_s \cdot C_{bl} = 4\cdot10^{-9}$ s. The maximum current of discharge $I_{max} = Q/\tau_{bl} = 5$ kA. A depth of a current penetration into a conductor (a thickness of a skin-layer) is $\delta = 50.33\ (\rho\,\tau)^{1/2}$ cm. (Here ρ is a specific resistance of a conductor, Ohm·mm²/m, and τ is a pulse duration). For copper $\rho_{Cu} = 1.75\cdot10^{-2}$ Ohm·mm²/m and at $\tau = 4\cdot10^{-9}$ s a thickness of the skin-layer is $\delta_{Cu} = 4\cdot10^{-4}$ cm. Assuming that the specific resistance of a person's skin is greater than the specific resistance of copper by 10^3 times, we find the thickness of the skin-layer at BL discharge $\delta_{bl} = 1.3\cdot10^{-2}$ cm. Discharge duration of typical linear lightning is about 100 μs. For the pulse of such duration a depth of a current penetration in the person's body is $\delta_{ll} = 2$ cm [26]. It is known, that danger of death is produced by passage of the current of 40-60 mA through a heart or the respiratory centre of brain [19, pp 230]. It is possible to expect, that a current of BL discharge, most likely, will pass through a skin of the victim, not coming in the vital centers inside a body. With this respect a BL blow is less dangerous, than a blow of a linear lightning [5].

An insulator between BL core and its cover is non-ideal. Therefore it is quite possible that in the over-breakdown electric field near BL surface a corona discharge, fed by a current of its charge leaking, can arise. Formation of ozone and nitrogen dioxide is probable in this discharge. Presence of these gases in the BL trace was experimentally proved by M.T. Dmitriev. He managed to observe BL on the bank of the river Onega in 1965. It looked as a bright white-yellow ball of 6-8 cm diameter, surrounded by two covers: an internal dark-violet cover with a thickness of 1-2 cm and a bright-blue external cover with a thickness of 2 cm. It is quite probable, that the luminescence of these covers was caused by the corona discharge. Dmitriev had gathered air from the BL trace inside some evacuated flasks. "The analysis of the gathered gas has shown presence of ozone and nitrogen dioxide in quantities, appreciably exceeding their normal concentration in air. The concentrations of other gases, which are usually present in air, did not change" [25, 27].

5. Principles of approaching to explanation of ball lightning high energy content

So, BL ability to possess energy with density of (1-3)·1010 J/m3 is the experimentally proved fact, which requires an explanation. Furthermore, a solution of the problem of large energy amount accumulation inside the limited volume can play a key role in solving of the BL problem as a whole. Almost two centuries of searches for solution of BL mystery show that ignoring of this property and attempts to develop a theory of an object with the energy density, which is three orders of magnitude smaller than that, which was measured, lead investigations of this problem into the deadlock [4]. Attempts to reduce this problem to existence of some mysterious reservoir of energy feeding BL through invisible channels proved to be fruitless [28, 29, 30]. Proposals to consider that BL energy is stored inside a radioactive nuclei or it has the source in merge reaction of deuterium nuclei are also far from reality [2, 31, 32]. In our opinion, one can consider as the most realistic only models, which assume a kinetic energy of BL substance as the basic reservoir of its energy [3]. However, in this case a serious problem of creating of forces, confining a system of moving particles inside a restricted region of space, arises.

In the model of BL, developed by the authors of this article [3, 4, 9, 11, 26, 33, 34], along with a possibility of accumulation of huge amount of energy, as its basic property, presence of non-compensated electric charge in it is considered. The requirement of presence of the positive charge automatically arises at development of the dynamic electric condenser model – a basic element of BL core [26, 33]. At first sight, presence of non-compensated charge only complicates a problem of the forces search keeping together a compact energy rich system – BL. However, this is not so. Really, let us consider a hollow sphere with an internal radius R and with a cover with a thickness a. Let us place inside the sphere a considerable quantity of charged particles with the total charge Q. We consider that the cover does not let the charges to pass through it. If all charged elements are distributed on an internal surface of a cover in regular intervals, they will stretch it with a force [4, 9, 26, 33]:

$$F_Q = \frac{Q^2}{8\pi\varepsilon_0 R^2}. \tag{6}$$

Here ε0 is an electric constant, ε0 = 8,85·10-12 F/m. If the charges are moving inside the cover, instead of "sitting" motionlessly on the wall, then to the force FQ of "Coulomb" pressure on the wall a force of "magnetic" pressure of currents Fm will be added. This force is approximately equal to FQ, so the resultant force Fem, stretching the cover, becomes equal to:

$$F_{em} = \frac{kQ^2}{8\pi\varepsilon_0 R^2}. \tag{7}$$

Here k ≈ 2. Now let us discuss, what will occur with the cover if to place the electric charge Q inside the sphere. We will consider that the cover consists of dielectric, for example, water molecules, and charge Q is placed in the sphere center. In the electric field E = Q/(4πε0R2), produced by this charge, there will be a polarization of water molecules: on the cover's outer side positive ends of dipoles, and on internal –negative ends will be located. The distance between water molecules in an ice crystal is equal to 3·10-10 m and on the area of 1 m2 1019 molecules will be located. If to accept a charge of the end of a molecule equal to the electron charge e = 1.6·10-19 C, then the density of charges on a cover surface is equal to σ = 1.6 C/m2. The chain of the polarized molecules of water, which length is equal to a thickness of a cover a, represents an electric dipole with the moment d = ea.

Let us assume that the charge Q is positive. Then the inner surface of the cover will be dotted with negatively charged ends of molecules. These charges are attracted to the central charge Q with a force

$$F_+ = \frac{4\pi R^2 \sigma Q}{4\pi\varepsilon_0 R^2}. \tag{8}$$

Here 4π R2σ is the total charge of the dipoles' ends on the inner surface of the cover. The positively charged ends of the dipoles are located on the outer surface of the cover. The number of these ends is equal to the number of the dipoles on the inner surface of the cover. The outer surface of the cover is repelled from the charge Q with a force

$$F_- = \frac{4\pi R^2 \sigma Q}{4\pi\varepsilon_0 (R+a)^2}. \tag{9}$$

The resulting force, shrinking the cover, is

$$F_{sh} = F_+ - F_- = \frac{Q\sigma a(2R+a)}{\varepsilon_0 (R+a)^2}. \tag{10}$$

We can see that the shrinking force Fsh = A(R,a)·Q is proportional to the first degree of a charge, and the force Fem = B(R)·Q2, stretching the cover, is proportional to a charge square. Therefore in the range of small values of a charge (from Q = 0 to Q = Qmax) conditions, when the shrinking force Fsh will be more than the stretching force Fem , can always be created.

In Fig. 4 plots of forces Fem and Fsh dependence on the charge Q, acting on a sphere cover with internal diameter 2R = 10 cm and thickness of the wall a = 1 cm, are shown. It is possible to see that up to Q = 1.53·10-2 C the force of the cover compression exceeds the force of its stretching. At the values of the charge, smaller than limit, the equilibrium condition of system can be supported due to an additional pressure upon an internal surface of the cover, created by the charge carriers, possessing kinetic energy.

Let the cover is stretched by a centrifugal force of objects with a total mass M, which glide with a velocity V on the inner surface of the cover. In this case a force Fk, stretching the cover due to particles movement is:

$$F_k = \frac{MV^2}{R} = \frac{2E_k}{R}, \tag{11}$$

where E_k is the total kinetic energy of the particles.

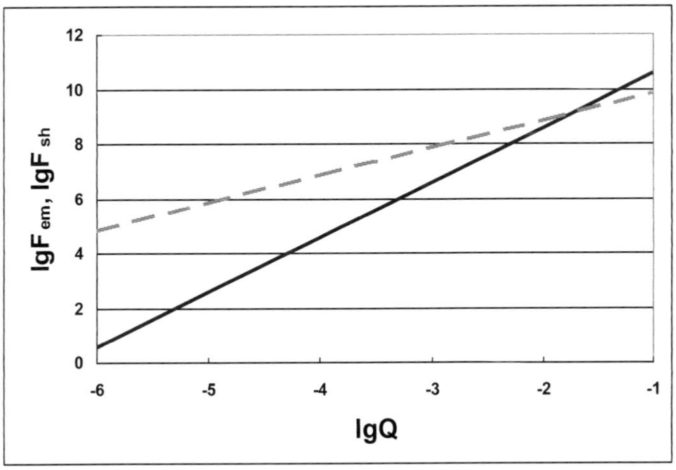

Figure 4. Lg Fsh (dotted line) and lg Fem (solid line) versus lg Q at R = 5 cm and a = 1 cm. Forces Fsh and Fem are in Newtons, charge Q is in Coulombs.

One can find that the same expression for F_k is obtained if particles with a total kinetic energy Ek move chaotically. According to Fig. 4, for R = 5·10-2 m and a = 10-2 m at Q = 10^{-2} C Fk = Fsh – Fem = 2.97·10^8 N. A volume of a sphere with internal radius 5 cm and wall thickness 1 cm is equal to V = 0.9·10^{-3} m3. According to formula (11), Ek = (Fk·R)/2. Thus, at Q = 10^{-2} C kinetic energy Ek = 7.42·10^6 J and energy density ρW =8.25·109 J/m3. As we see, the density of kinetic energy of particles at Q = 10^{-2} C comes nearer to the value of energy density of highly energetic BL (ρ= 10^{10} J/m^3).

Above we already paid attention to that even for BL, possessing a charge Q = 2·10-5 C, an electric field intensity near its surface appears to be above the breakdown value Ebr = 3·10^6 V/m by 40 times. It is especially fair for the considered BL with the charge of Q = 10^{-2} C, for which the field strength is E = 2.5·10^{10} V/m and potential U = Q·R = 1500 MV at R = 6·10^{-2} m. The engineers, testing high-voltage devices know that already at potentials 2 - 5 MV between the electrodes, displaced at some tens of meters, a spark discharge can pass [35]. The huge internal resistance of the generator of a current – the charged core of BL – can be a reason of such discharge absence between BL and subjects surrounding it. It is known, that a way, which the spark discharge should pass, is laid by a leader – a thin channel with a rather high temperature 3000-6000 K [36]. For heat maintenance (and, means, for high conductivity) of the leader during time of its "germination" (10-4 - 10-3 s) it is necessary that through the channel the current, extracting a power not less than 130 W for centimeter of its length, was flowing. At smaller powers the channel cools down, its conductivity decreases, and advancement of the leader stops. According to estimations, for typical cases of a spark development in air at atmospheric pressure the current flowing through the leader channel cannot be smaller than 1 A [35, 36 (52, 53)]. However, for BL because of low conductivity of an interval between the charged core and its cover the current is limited by values of 10^{-3} - 10^{-2} A. This cannot become an obstacle for maintenance of corona discharge on its surface, but, suppresses the possibility of spark channel formation [22]. It is fair for the "healthy" BL, which is in an equilibrium condition. At isolation infringement between a core and a cover BL charge can be discharged on surrounding subjects through the spark channel. Such cases, really, were observed [5, 24, 10, 25].

IOP Conf. Series: Journal of Physics: Conf. Series **996** (2018) 012011 doi:10.1088/1742-6596/996/1/012011

6. Details of heterogeneous models of ball lightning

In the previous sections we defined general requirements for BL structure. We came to conclusion that conditions of accumulation of large density of kinetic energy can be created in the two-component system, consisting of the charged core and the dielectric cover. Now it is the time to discuss in more detail a structure of the BL core and to discuss how the definite model can explain observable properties of BL. We will present here two approaches to creating of BL core model. According to the classification, used in [3], we name these models by "Model N" and "Model B".

1. Model N.

In this model it is supposed that the space inside the BL cover is filled with certain "plasmoids", possessing a non-compensated electric charge. The plasmoid is meant as a system of moving charges, kept together by own electric and magnetic fields [37]. However, it was proved that stabilization of the plasmoid only by internal forces is, basically, impossible. Any external force, for example, a force of gravitation, is necessary for this purpose [38]. In our case a role of this external force is played by the force of the cover compression. According to the formula (11) for BL with internal radius of the cover $R = 5{\cdot}10^{-2}$ m at charge value $Q = 10^{-2}$ C the force, compressing the cover, is $F_k = 2.97{\cdot}10^{8}$ N, and a pressure inside the cover is $P_{in} = F_k/(4\pi R^2) = 9.5{\cdot}10^{9}$ Pa. That is almost by 10^5 times greater than a force of the atmospheric pressure.

As a prototype of our plasmoid "the hydromagnetic plasma condenser", invented in the fifties of 20th century, serves [39]. This condenser represents a disk of plasma, placed in a strong magnetic field. In this field positive and negative charges are separated: electrons move in an orbit of small radius around a magnetic field direction, and positive ions move in an orbit of the big radius. Movement of charges influences not only the external magnetic field, but also the electric field between regions of separated charges. Basically, for existence of similar system of moving charges presence of an external magnetic field is not obligatory. Really, let us imagine a certain ring of electrons, moving in a closed orbit of radius r in magnetic field H, which is created by the positive ions, rotating in an orbit of big radius R. These electrons are kept in the orbit, making drift movement in mutually perpendicular fields H and E, where E is an intensity of electric field between the internal electronic ring and the external ionic ring. Ions are kept in the orbit thanks to their attraction to the electronic ring (due to action of the field E on them). Such a configuration is named in [26] "the dynamic electric condenser", and the concept – the "electrodynamic model of BL" [33 (49)].

Equating a centripetal force $F_e = eE$ to centrifugal force $F_c = (m_p v_p)^2/R$, we find an orbital velocity of an ion $v_p = (eER/m_p)^{1/2}$. Here m_p is a mass of proton (of the most light positive ion), $m_p = 1.67{\cdot}10^{-24}$ kg; $e = 1.6{\cdot}10^{-19}$ C is a proton charge. We consider not relativistic case, therefore an intensity of electric field at $R = 10^{-6}$ m should not exceed $E = 10^{14}$ V/m, and at $R = 10^{-2}$ m must be $E < 10^{10}$ V/m. Rotation frequency of proton is $f_p = v_p/(2\pi R) = (eE/(4\pi^2 m_p R))^{1/2}$. If the total charge of all protons is equal to Q_p, a current I_p, created by their movement, is $I_p = Q_p f_p = (Q_p^2 eE/(4\pi^2 m_p R))^{1/2}$. This current generates a magnetic field intensity in the center of the proton orbit $H_p = I_p/2R = (Q_p^2 eE/16\pi^2 m_p R^3)$. As rough approach let us accept that electric field E, created by an electronic ring with the charge Q_e, is equal to a field of a point charge. (More exact expression can be found in [26, 33]). In this case

$$E = Q_e/(4\pi\varepsilon_0 R^2),\ \text{V/m}. \quad (12)$$

(Here $\varepsilon_0 = 8.85{\cdot}10^{-12}$ F/m is the electric constant). For such E

$$H_p = (Q_p^2 e Q_e/(64\pi^3 m_p \varepsilon_0 R^5))^{1/2}. \quad (13)$$

Movement of electrons in mutually perpendicular electric E and magnetic H fields occurs in a cycloid [40)]. For the way of electron was closed, it is necessary that the magnetic field was strong enough and the condition $H_p \geq (\varepsilon_0/\mu_0)^{1/2} E$ was satisfied [41)]. Here $\mu_0 = 1.25{\cdot}10^{-6}$ V·s/A·m is the magnetic constant. In this case we find from (12) and (13) the relation between values of total charges of protons Q_p and electrons Q_e:

$$Q_p^2 \geq (4\pi m_p R Q_e)/e\mu_0 = 0.1R\, Q_e. \quad (14)$$

The analysis shows, that at $Q_e = 10^{-6}$ C the inequality (14) is carried out at $R > 10^{-4}$ m, and at $Q_e = 10^{-3}$ C, for its performance it is necessary, that there was $R > 10^{-2}$ m. It means that in this area of values Q_e and R the total charge of protons exceeds the total charge of electrons, so the plasmoid (the dynamic electric condenser) is positively charged. For a movement of the charged particles without being braked due to collisions with air molecules, a high vacuum should exist inside the cover. Simultaneously the vacuum plays a role of an electric insulator between the dynamic condenser and the BL cover. The radius of the proton orbit aspires to increase. The expansion of the orbit can be constrained by mutual impacts of dynamic condensers (if there are many condensers inside the cover) or by their impacts with the cover wall. Due to losing of the core charge the cover becomes positively charged, this promotes process of "plasmoids" reflection from its wall.

In the considered above example we spoke about properties of BL of an average size with the diameter $2(R + a) = 2(5 \text{ cm} + 1 \text{ cm}) = 12$ cm. Now let us discuss how its characteristics will vary at reduction of a size of its elements. Kinetic energy of the proton, hold in an orbit by field E, created by the electronic ring (see formula (12)), is $w_p = m_p v_p^2/2 = eQ_e/(8\pi\varepsilon_0 R)$, and the total kinetic energy N_p of protons ($N_p = Q_p/e$) is:

$$W_p = w_p \cdot N_p = Q_e Q_p/(8\pi\varepsilon_0 R) = 4.5 \cdot 10^9\, Q_e Q_p/R. \tag{15}$$

Considering a relation between values of total charges of protons and electrons (14), the formula (15) can be written in a form:

$$W_p \le \left(\frac{m_p}{16\pi\varepsilon_0^2 e\mu_0} \right)^{1/2} \cdot \frac{Q_e^{3/2}}{R^{1/2}} = 1.45 \cdot 10^9 \cdot \frac{Q_e^{3/2}}{R^{1/2}} \text{ (J)}. \tag{16}$$

Accepting the plasmoid volume equal to $V_c = 4\pi R^3/3$, we find an average density of kinetic energy of protons: $\rho_w = W_p/V_c = 0.35 \cdot 10^9 \,(Q_e^{3/2}/R^{7/2})$ (J/m^3). We see that at decrease of the plasmoid radius R the density of total kinetic energy grows. The limit to decreasing of the size R puts the requirement that the number of electrons in the system was not less than 100 ($Q_e \ge 1.6 \cdot 10^{-17}$ C). Besides, a cover of dielectric, capable to constrain an expansion of a core of the tiny ball lightning must exist. Let us estimate the minimum size of BL with a total charge of electrons $Q_e = 1.6 \cdot 10^{-17}$ C.

Let in the formulas (7), (10) and (11) $k = 2$, $E_k = W_p$. At small R the balance equation of forces, acting the shell of the plasmoid, can be written as:

$$\frac{2W_p}{R} + \frac{Q^2}{4\pi\varepsilon_0 R^2} - \frac{Q\sigma a(2R + a)}{\varepsilon_0 (R + a)^2} = 0. \tag{17}$$

Because $R >> a$ let's disregard terms with a^2. In the result we have:

$$\frac{2W_p}{R} + \frac{Q^2}{4\pi\varepsilon_0 R^2} = \frac{4Q\sigma a}{\varepsilon_0 (R + 2a)}. \tag{18}$$

Solving this equation with respect to a, we find:

$$a = \frac{RX}{1 - 2X}, \text{ where } X \equiv \frac{\varepsilon_0}{4\sigma Q} \left(\frac{2W_p}{R} + \frac{Q^2}{4\pi\varepsilon_0 R^2} \right). \tag{19}$$

Accepting the charge Q equal to the total charge of protons Q_p, σ=1.6 C/m^2 and using expressions (14) and (16), we obtain:

$Q \approx Q_p = (0.1RQ_e)^{1/2} = 0.316\, R^{1/2} Q_e^{1/2}$ and

$$X = \frac{12{,}696 \cdot 10^{-3} Q_e}{R^2} + \frac{3{,}937 \cdot 10^{-3} Q_e^{1/2}}{R^{3/2}}. \tag{20}$$

According to (19) the condition $X < 1/2$ should be satisfied. At small R and Q_e the first term in the expression (20) is much smaller than the second, therefore it is possible to write $R^3 \ge (7.87 \cdot 10^{-3})^2 Q_e$. From here we find the minimum internal radius of the BL cover with a charge Q_e: $R \ge 3.96 \cdot 10^{-2}\, Q_e^{1/3}$. At $Q_e = 1.6 \cdot 10^{-17}$ C radius $R \ge 100$ nanometers. Estimations by the formula (19) show that the thickness a of a cover of tiny BL with the total charge of electrons $Q_e = 1.6 \cdot 10^{-17}$ C at increase R from

75 to 150 nanometers decreases from 493 nanometers to 28 nanometers and its diameter $2(R + a)$ becomes about 360 nanometers. Apparently, it is the minimum size which BL can have. Let the cover consists of water and its thickness is equal to 28 nanometers. The size of water molecule is $r_w = 1.5 \cdot 10^{-10}$ m, hence, a cover can consist of $n_s = a/r_w = 190$ layers of molecules. Kinetic energy of ions in such BL (see the formula (16)) is $W_p = 1.45 \cdot 10^9 Q_e^{3/2}/R^{1/2} = 2.4 \cdot 10^{-13}$ J $= 1.5 \cdot 10^6$ eV. Having divided W_p on volume of BL with the cover, $V_{bs} = 4\pi (R+a)^3/3 = 2.36 \cdot 10^{-20}$ m^3, we find the energy density $\rho_{ws} = 10^7$ J/m^3. This value is much smaller, than the limiting energy density of BL. At growth of the total charge of electrons Q_e and at increase of BL size the density of protons kinetic energy, counting on total volume of BL V_{bs}, can considerably exceed the above found figure. For example, for BL with internal radius of the cover $R = 1$ mm and the charge $Q_e = 10^{-6}$ C its diameter will be $2(R+a) = 2.64$ mm, kinetic energy of protons W_p will reach 46 J, and the energy density – $\rho_{ws} = 5 \cdot 10^9$ J/m^3. A quantity of these small BL can appear inside a big container with a dielectric wall. If a total volume of such ball is equal to 1 liter (it is a typical volume of "macroscopic" BL), than their total energy will be nearby 5 MJ. If the cover of this container – a "composite" BL – appears to be transparent, then these small BL can be seen with naked eyes [20].

One of basic properties of BL is its ability to soar freely in air it marked in [5, 24, 10, 20, 25, 42]. Let us check, whether a considered model of BL in the form of hollow sphere with internal radius $R =$ 5 cm and a cover thickness $a = 1$ cm is capable to such action. We will assume that the cover consists of the substance, which density is close to density of liquid water $\rho_w = 10^3$ kg/m^3. Volume of the cover of our BL is $V_{sh} = 4\pi[(R+a)^3 - R^3]/3 = 3.8 \cdot 10^{-4}$ m^3, and its mass $m_{sh} = \rho_w \cdot V_{sh} = 0.38$ kg. Weight of the cover is $P_{sh} = m_{sh} \cdot g = 3.74$ N. Let an area inside the cover is filled with carriers of the positive charge, which total value is $Q = 10^{-3}$- 10^{-2} C. We will assume that charge carriers are protons. In this case their quantity in the cover is $N_p = Q/e = 6.25 \cdot 10^{15} - 6.25 \cdot 10^{16}$, and a total weight $M_p = m_p \cdot N_p = 10^{-8}$-$10^{-7}$ kg. Near the ground surface an electric field always presents. At fine weather its intensity is nearby $E_{gw} =$ 100-150 V/m, and during thunder storms it can be greater than E_{gw} by 100-1000 times [43]. For creation of an elevating force, equal to the weight ($P_{sh} = 3.74$ N) of BL with charge $Q = 10^{-2}$ C, an electric field with intensity $E = P_{sh}/Q = 3.74 \cdot 10^2$ V/m is required. Because of leaking of BL charge into a space around it, a cloud of charges appears. Repelling of BL from this cloud can provide it soaring over conductors [5, 24, 7, 10, 20, 25, 42, 44)] and existence of groups of BL [46]. Elements of BL core represent magnetic dipoles. Because of chaotic movement of these elements inside a cover of BL, it has no macroscopic magnetic moment [34]. However, in an external magnetic field there can be a partial ordering of the moments of separate dipoles. Likely this can explain a case of lifting of horseshoe-shaped magnets by BL in air [7].

As the generator of electromagnetic radiation of BL electric charges (electrons and protons), moving on the closed orbits, apparently, serve. At performance of condition $H_p \geq (\varepsilon_0/\mu_0)^{1/2} E$ a velocity of electrons appears to be close to the light velocity. Such electrons become a source of a synchrotron radiation which frequency appears n times greater than the frequency of their orbital movement $f_e = c/(2\pi r)$ (r – is a radius of an electronic orbit). A harmonic number n_{max} on which a maximum of radiation intensity falls, is [47]:

$$n_{max} = (3/2) \cdot (1 - \beta^2)^{-3/2} \qquad (21)$$

Here $\beta = v_e/c$ – is a ratio of electron velocity to the light velocity. A power of synchrotron radiation is inversely proportional to product of squares of radius of the electron orbit r and a parameter $\alpha = 1 - \beta$ [47]. Calculations show, that if there is a "mechanism" of protons kinetic energy transfer to electrons, then a full energy of BL should "be highlighted" by electrons during 10^{-3}-10^{-2} s [33]. However, this conclusion concerns only a case of light radiation by a single electron or by an electron bunch. If electrons are located along an orbit absolutely in regular intervals, they will not radiate light [40]. In our system with large number of electrons they will aspire to be distributed in an orbit in regular intervals. The wave length, on which a maximum of radiation intensity falls, depends on electrons velocity and can be different. Therefore a color of BL can be any. At infringement of uniformity of electrons arrangement along an orbit, BL can radiate energy in a form of a short pulse. Apparently, the specified mechanism of BL radiation is realized not only in optical range, but also in the radio range.

The most probable reason of BL formation is a discharge of linear lightning (or, strictly speaking, any channel of enough strong current). Usually a voltage between a storm cloud and the earth is $U = 10^9$ V. A discharge of average lightning transfers a charge $Q_l = 5$ C to the earth, from here its energy is $W_l = U \cdot Q_l = 5 \cdot 10^9$ J. At the energy density $\rho_{bl} = 10^{10}$ J/m^3 BL with a volume $V_{bl} = 10^{-3}$ m^3 possesses energy 10^7 J, that is 0.2 % of the linear lightning energy. According to the "scenario", described in [3, 33, 34, 48], BL is generated near a bend of the linear lightning channel at passage of several return strokes through it. It is considered that thanks to action of the first return stroke in a space near to the bent channel the vacuum cavity is formed and there air ionization takes place. At passage through the channel of a current of the next discharge, a pulse of magnetic field with duration of forward front nearby 10 μs and with duration of a falling down part nearby 50 μs is generated. In a phase of increase of the magnetic field eddy electric field is generated. Under the influence of this field a separation of charges takes place: electrons are injected inside a vacuum cavity, and positive ions (protons) remain near its walls. During reduction of intensity of the magnetic field the direction of the eddy electric field vector changes to reverse. As a result of action of electric and magnetic fields, the acceleration of protons takes place, and a part of electrons, moving along magnetic field lines leaves the area near the channel of the lightning. As a result a positively charged system of moving charges (the dynamic electric condenser) is formed: an electronic ring, round which protons rotate. In electric field of non-compensated charge from water drops a cover can be formed. The single double ring of moving charges is thermodynamically unstable: according to the law of growth of entropy it is favorable for it to divide into elements of small size, which will fill in regular intervals a space inside the cover. The minimum size of these elements can be about 1 mm [11, 33, 34].

2. Model B.

Another "material" model of BL is the "chemical-thermal" model [11, 49, 50, 51, 52]. According to it, BL is a unipolarly charged object with a shell and a nucleus. The core can consist of atomized particles of water vapor or atoms of matter that appear when a linear lightning strikes a variety of materials, such as SiO_2, Al_2O_3, etc., or the soil, in the presence of water in the soil or without it. For simplicity of analysis, we assume that the energy of linear lightning in this case is sufficient to transfer the material to which it was exposed, into atoms of an ideal gas. Such an approach makes it possible to consider the possibility of the formation of high-energy BL, although it leaves the question open about the properties of BL in the composition of which are nucleus of vapor and droplets of the formed matter.

Electrons carried by linear lightning, when struck in materials at atmospheric pressure, quickly adhere to neutral molecules, forming negative ions. Therefore, the core material contains these ions, which determines the electrical properties of BL. In this case, the shell is a thin film of a melt or solidified substance. The charge is transmitted to BL from a linear lightning, and the stability of BL with respect to destruction, as in [9, 19, 26], and is provided by the polarization force that arises from the Coulomb influence of a unipolar charge. Inside such BL there is a mixture of a gas and negative ions of the atomized substance.

In contrast to Model N, in this model there are components inside the shell that are associated with the random motion of particles inside the shell. In this case, the energy of such BL is determined mainly by the energy of evaporation evolved at the destruction of the object, to a lesser extent – when the substance of BL burns and the average energy of the particles inside the nucleus at a given temperature and pressure. The release of energy during combustion is insignificant with respect to evaporation. The calculation of combustion processes of BL material in the air requires a more complicated model, but does not make a significant contribution to the energy density of BL [49].

Energy of BL. The most important question in this approach is the question of the BL energy. In order to find an answer to this question, we first determine a gas pressure inside BL, acting its surface. Then, knowing the pressure and enthalpy of vaporization, that is, the energy density per atom of the substance or molecule of water, we will determine the specific energy of such ball lightning. We consider two possible cases of BL formation.

First, the composition of the shell and the gas inside BL differs little (experiments on the evaporation of basalt and steel [50-52] indicated this possibility). In this case, the existence of a shell is determined by the nonequilibrium processes in the plasma that occur when a linear lightning strikes the ground, when nonequilibrium processes of shell formation lead to a change in its physical properties, without affecting the chemical. For estimates in this case, silicon is considered, whose content in the earth's crust reaches 50%. The gas consists of SiO_2 silicon, and the shell consists of a modified melt of SiO_2 and its alloy with soil components. For estimates, we assume that the temperature of a gas inside the shell is close to 2000 K and is equal to typical boiling point of SiO_2.

Second: the core of BL consists of water vapor, and the shell – of SiO_2. For estimates, we also assume that the temperature of the gas inside the shell is of the order of 2000 K, i.e. of the order of the boiling point of SiO_2.

Let us write an equation of balance of forces acting the shell of such BL:

$$F_{surf} + F_a + F_{sh} - F_{el} - w4\pi(R+a) = 0, \tag{22}$$

here $F_a = 4\pi(R+a)^2 P_a$ – is a force of atmospheric pressure, $P_a = 1.03 \cdot 10^5$ Pa is the atmospheric pressure, R is BL radius and a is a thickness of BL shell.

$F_{sh} = \dfrac{\sigma a(2R+a)}{\varepsilon_0 (R+a)} Q$ is a force of a polarization interaction of charges [11(27)], σ is a surface density of charges on a shell (see above), Q is BL charge, $\varepsilon_0 = 8{,}85 \cdot 10^{-12}$ F/m is the electric constant. $F_{el} = \dfrac{1}{4\pi\varepsilon_0} \dfrac{Q^2}{(R+a)^2}$ is Coulomb interaction of charges of the same sign force, which influences the shell [11]. $F_{surf} = 8\pi\alpha(R+a)$ is the force of surface tension, α is a coefficient of surface tension of the shell, w – is the pressure inside BL (for simplified consideration, we do not take into account the own volume of the shell) – determines the internal energy of BL, which goes to collisions of gas particles with the BL shell and leads to its stretching. We believe that the processes of atomization of matter, Coulomb repulsion and polarization compression of the shell lead to the establishment of pressure inside BL. In its turn, the atmospheric pressure and surface tension practically do not contribute. From (22) follows the equation for its determination

$$w = \frac{F_{surf} + F_a + F_{sh} - F_{el}}{4\pi(R+a)^2} \approx \frac{F_{sh} - F_{el}}{4\pi(R+a)^2}. \tag{23}$$

A concentration of particles N_a inside ball lightning can be determined on a basis of the equation

$$w = P = N_a kT, \tag{24}$$

here $k = 1{,}38 \cdot 10^{-23}$ J/K is Boltzmann constant, T – the temperature of particles inside BL.

BL mass is determined as:

$$M = m_a N_a V = \frac{4}{3}\pi R^3 m_a N_a, \tag{25}$$

here m_a is a mass of a gas particle. BL energy density can be determined by an equation

$$E = \frac{w}{kT}\left(m_a \lambda + \frac{3}{2}kT\right), \tag{26}$$

λ is an enthalpy of a vapor – gas formation (energy of sublimation), or an energy density of water atoms recombination, (J/kg). At the main material SiO_2 $\lambda \approx 9.55 \cdot 10^6$ J/kg [17 (35)], and at the main material H_2O $\lambda \approx 5.1 \cdot 10^7$ J/kg (it is determined by data on water molecule bonds breaking [53]). In all the cases we have considered $m_a \lambda >> \frac{3}{2}kT$, which is natural, because the sublimation energy is of the order of the dissociation energy, and that, in turn, is much larger than the thermal energy of the gas particles.

On a graph of Fig. 5 one can see typical dependences of pressure w (J/m^3, its dimension coincides with Pa) over the charge Q of BL. The strength of BL shell polarization compression is proportional to the charge Q, and the Coulomb force resulting in the rupture of the shell is proportional to Q^2, they act in different directions, so there is a charge value at which these interactions are compared. With a larger value of BL charge, its main energy goes to stretching of the shell, and the internal energy that can be released and do a work falls. From Fig. 5 one can see that the pressure $w(Q)$ varies widely depending on the value of Q and has a maximum at given values of the shell thickness and the density of the polarization charges located on it.

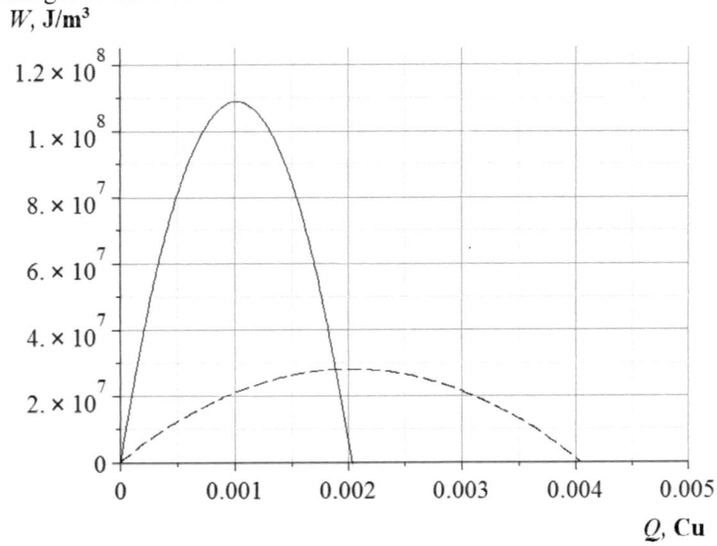

Figure 5. Typical values of w (J/m^3) with respect to BL charge Q. Solid line $R = 5$ cm, dashed line $R = 10$ cm, $\sigma = 1.5$ C/m^2, $a = 1$ mm.

Table 1 a)-b) presents parameters of BL for various values of (σ, a, R, Q) calculated by formulas (22) - (26) at the maximum value of w (Q) which we are interested in, since we analyze an existence possibility of high-energy BL. The data in Table 1 show that the BL energy density varies widely, mainly depends on the shell material, which determines the surface density of the charge, the charge inside BL, a thickness of the shell, and a radius of the BL. The energy density depends little on a type of material inside the shell. Within the framework of this model, BL can have energy comparable with the observed.

Table 1 a). BL parameters for determination of pressure w.

N	σ, C/m^2	a, mm	R, cm	w, J/m^3	Q, C
1	1.5	1	5	$9.60 \cdot 10^7$	0.001
2	1.5	1	10	$2.45 \cdot 10^7$	0.002
3	1.5	1	50	$1.00 \cdot 10^6$	0.01
4	4.5	1	5	$8.59 \cdot 10^8$	0.003
5	4.5	1	10	$2.22 \cdot 10^8$	0.006
6	4.5	1	50	$9.06 \cdot 10^6$	0.029
7	4.5	10	5	$5.28 \cdot 10^{10}$	0.03
8	4.5	10	10	$1.72 \cdot 10^{10}$	0.06
9	4.5	10	50	$8.52 \cdot 10^8$	0.29

Table 1 b). Parameters for determination of BL energy density.

N	n, m^{-3} SiO_2, H_2O	M,kg SiO_2	M, kg H_2O	E, J/m^3 , SiO_2	E, J/m^3, H_2O
1	$3.46\cdot 10^{27}$	0.19	0.06	$3.19\cdot 10^{9}$	$5.10\cdot 10^{9}$
2	$8.91\cdot 10^{26}$	0.37	0.11	$8.16\cdot 10^{8}$	$1.31\cdot 10^{9}$
3	$3.64\cdot 10^{25}$	1.83	0.55	$3.32\cdot 10^{7}$	$5.30\cdot 10^{7}$
4	$3.11\cdot 10^{28}$	1.67	0.5	$2.86\cdot 10^{10}$	$4.57\cdot 10^{10}$
5	$8.02\cdot 10^{27}$	3.33	1.0	$7.38\ 10^{9}$	$1.18\ 10^{10}$
6	$3.29\cdot 10^{26}$	16.6	5.0	$3.01\cdot 10^{8}$	$4.82\cdot 10^{8}$
7	$1.93\cdot 10^{30}$	166	50	$1.76\cdot 10^{12}$	$2.82\cdot 10^{12}$
8	$6.24\cdot 10^{29}$	333	100	$5.73\cdot 10^{11}$	$9.18\cdot 10^{11}$
9	$3.12\cdot 10^{28}$	1650	493	$2.83\cdot 10^{10}$	$4.52\cdot 10^{10}$

Radiation of BL. A question of BL radiation reason in this model does not have an unambiguous answer, since BL has complex internal and surface structure.

The cause of BL radiation generation in the optical and radio ranges can be a plasma created by an electric field of a charge, a high core temperature, the movement of electrical charges in the plasmoids that form its core, or processes that cause the plasma pulsations near its shell. Some of these issues were discussed in [19, 22].

The radiation of BL can be of two types. It can be equilibrium then the BL shell is illuminated, filled with hot vapor. In this case, the heat is transferred to the shell by particles of vapor inside the shell.

The radiation of BL can be also a nonequilibrium one. Then under the action of BL large charge, a plasma appears in the strong electric field near BL surface as a result of ionization processes. The lifetime of the nonequilibrium radiation is determined by a plasma lifetime at its surface, i.e. the time of the charge decreases. Parameters of the nonequilibrium radiation require a special analysis with development of this model, which is beyond the scope of this article.

The parameters of the equilibrium radiation can be estimated as follows. We assume that the heating of the shell by the vapor is compensated by thermal radiation from the surface of the shell, then we can use the following expression for the specific heat of a real gas [54]

$$\frac{3}{2}Nk_B\frac{4}{3}\pi R^3\frac{dT}{dt}=-\sigma_{SB}T^4 4\pi R^2, \tag{27}$$

where k_B is the Boltzmann constant, and σ_{SB} is the Stefan-Boltzmann constant. The solution of this equation for determining of the cooling time to the temperature T_1, if the initial gas temperature is much greater than the final gas temperature $T >> T_1$, has a form:

$$t_x=\frac{Nk_BR}{6\sigma_{SB}T_1^3}. \tag{28}$$

For example, let us substitute the following values of BL average parameters from Table 1 into (28): $N = 8{\cdot}10^{27}$ m^{-3}, $R = 0.1$ m, $T_1 = 500$ K, we get $t_x \approx 260$ s. This value is comparable to the observable time of BL.

In works [55, 56] in erosive discharges on the basis of polymeric materials, luminous spherical objects up to 2 cm in diameter were obtained, which emitted glow for 1-2 seconds. At the same time, it seemed that the intensity of their radiation did not change during their lifetime, and then these objects instantly extinguished. Analysis of the video shooting showed that during 0.04 seconds there was a smooth fading, which was not recorded by an eye of the observer. Therefore, a constant intensity of the luminescence noted by BL observers is most likely an optical effect due to the fact that the eye does not distinguish the decrease in the intensity of the glow when BL is cooled from 2000 K to ~ 800 K. Subsequent cooling of BL, and a corresponding decrease in the luminescence intensity occurs rapidly. During this time of extinction, the eye does not have time to adapt itself to a low intensity of the luminescence, since the physiological adaptation time of the eye when observing bright objects is ~ 10 s [57].

According to (28), the emission time of BL radiation increases with the radius of the ball, so the increase in the lifetime of BL with an increase in its radius, observed by observers [58], is easily explained. The color of BL in this model is determined by a material of which its shell consists. Being movable by the electric field of the earth, ball lightning can approach a window glass and "penetrate" into the room through a crack or a crackle [20]. In addition, it is capable, having spent part of its energy, to heat the glass to a melting temperature and form a hole in it [59]. But this "material" BL is unable to pass through a glass leaving no traces at all.

We described some of the features of this model. At present, the problem of BL transformation from a tape into a ball, and also folding a tape into a sphere, remains an open question in the framework of this model, the question of the magnetic properties of such BL is open, etc. However, there are reasons to suppose that these questions will find answers within the framework of classical mechanics and electrodynamics.

7. Conclusion

We have chosen here three "the link main things" of BL problem: high energy storage, presence of non-compensated electric charge and dynamic character of its arrangement. We came to conclusion, that these properties can be combined in the system, consisting of three elements: unipolar charge, a dielectric cover and charge carriers, which simultaneously serve as accumulators of BL energy. This conclusion, along with conclusion about "material" nature of BL, which we made on the basis of analysis of its interaction with glasses [60 (4)], allows to one, creating a theory of BL, to narrow a number of its models, having excluded from them these which cannot explain high energy storage of BL and its ability to possess non-compensated electric charge.

8. Acknowledgement

We thank Dr. V.I. Kulikov for useful discussion of the results of ball lightning explosion.

References

[1] Arago F 1838 *Sur le Tonnerre.* Annuaire au Roi par le Bureau des Longitudes. Notices Scientifiques pp 221

[2] Nikitin AI 2006 Will be a success of ball lightning problem solution in 21st century? *Khimicheskaya Fizika.* 25(3) pp 18-37. (in Russian)

[3] Nikitin AI 2004 The principles of developing the ball lightning theory. *J.Russian Laser Research* 25(2) pp 169-191

[4] Nikitin AI 2012 A new approach for solving of ball lightning problem – change of the paradigm *Vestnik RAEN* 2 pp 19-30. (in Russian)

[5] Stenhoff M 1999 *Ball Lightning. An Unsolved Problem in Atmospheric Physics.* New York: Kluwer/Plenum

[6] Goodlet BL 1937 Lightning *IEE J. London* 81 pp 1
[7] Imyanitov I, Tikhiy D 1980 *Beyond a Boundary of Science Laws* Moscow: Atomizdat, (in Russian)
[8] Egely G 1993 Analysis of Hungarian ball lightning observations *In: Progress in ball lightning research* Ed. AG Keul. Proc. VIZOTUM. Salzburg, Austria
[9] Nikitin AI, Bychkov VL, Nikitina TF, Velichko AM 2014 High-energy ball lightning observations *IEEE Transactions on Plasma Science* 42(12) pp 3906-3911
[10] Brand W 1923 *Der Kugelblitz* Hamburg: H. Grand
[11] Bychkov VL, Nikitin AI, Dijkhuis GC 2010 Ball lightning investigations *In: The Atmosphere and Ionosphere. Dynamics, Processes and Monitoring.* Eds. VL Bychkov, GV Golubkov, AI Nikitin. Dordrecht: Springer pp 201-373
[12] Dugarov BTs An accident in Mogsokhon http://arigus-tv.ru/news/item/50247/ (in Russian)
[13] *A Management on Blasting Works* Moscow: Military Publishing House of the USSR Ministry of Defence, 1963. (in Russian)
[14] *The Big Soviet Encyclopaedia* 1956 Ed. BL Vvedensky 43. Moscow: BSE, (in Russian)
[15] Likhoshyorstnykh GU 1994 Anomalous cases of meetings with ball lightning *In: Ball Lightning in Laboratory* Eds. RF Avramenko, VL Bychkov, AI Klimov, OA Sinkevich. Moscow: Khimia pp 218-223. (in Russian)
[16] Dmitriev MT, Bakhtin BI, Martynov VI 1981 Investigation of thermal factor of ball lightning *J. Tekhn. Fiz.* 51(12) pp 2567-2572 (in Russian)
[17] *Physical Values. Reference book.* Eds. IS Grigoriev, EZ Meylikhov. Moscow: Energoatomizdat, 1991 (in Russian)
[18] Batygin A, Mosin I 1989 Visit of "fiery" lady *Newspaper "Pravda"* N. 220. August, 8 (in Russian)
[19] Bychkov VL, Nikitin AI 2014 Ball lightning: A new step in understanding *In: The Atmosphere and Ionosphere. Elementary Processes, Monitoring, and Ball Lightning.* Eds. VL Bychkov, GV Golubkov, AI Nikitin Cham: Springer pp 201-367
[20] Grigoriev AI *Ball Lightning* Yaroslavl: YarGU Publisher, 2006 (in Russian)
[21] Landau LD, Lifshits EM 1959 *Electrodynamics of Condensed Media*. Moscow: Publishing of Phys.-Math. Literature (in Russian)
[22] Nikitin AI, Nikitina TF, Velichko AV 2013 Radio emission of corona discharge and ball lightning *Electrichestvo* 9 pp 12-22 (in Russian)
[23] Kalashnikov SG 1985 *Electricity* Moscow: Nauka (in Russian)
[24] Stakhanov IP 1996 *About Physical Nature of Ball Lightning* Moscow: Nauchny Mir (in Russian)
[25] Singer S 1971 *The Nature of Ball Lightning* New York: Plenum
[26] Nikitin AI 1998 Electrical capacitor as the element of the power core of ball lightning *Electr. Technol. Russia* 4 pp 70-85
[27] Dmitriev MT 1967 A nature of ball lightning *Priroda* 6 pp 98-106 (in Russian)
[28] Ohtsuki YH, Ofuruton H 1991 Plasma fireballs formed by microwave interference in air *Nature* 350(6314) pp 139-141
[29] Kapitsa PL 1987 About the nature of ball lightning *In: Experiment, Theory, Practice* Moscow: Nauka pp 55-61 (in Russian).
[30] Handel PH 1975 Maser theory of ball lightning Bull. *Am. Phys. Soc. Series ii.* 20(1) pp 26
[31] Altschuler MD, Houste LL, Hildner E 1970 Is ball lightning a nuclear phenomenon? *Nature* 228 pp 545-547
[32] Vlasov AN 2009 Is ball lightning an induction discharge in a vortical ring? *Nauka i Zhizn* 7 pp 28-34 (in Russian)
[33] Nikitin AI 2006 The electrodynamic model of ball lightning *Khimicheskaya Fizika* 25(3) pp 38-62 (in Russian)
[34] Nikitin AI, Velichko AM, Nikitina TF 2008 Principles of search of conditions of creating of ordered plasma structures *Proc. RAS. Energetics* 2 pp 115-132 (in Russian)
[35] Bazelian EM, Raizer Yu P 1997 *Spark Discharge* Moscow: MFTI Publishing (in Russian)

[36] Bazelian EM, Raizer Yu P 2001 *Physics of Lightning and Protection from Lightning* Moscow: FIZMATLIT (in Russian)
[37] Bostik WH 1956 Experimental study of ionized matter projected across a magnetic field *Phys. Rev*. 104(2) pp 292-299
[38] Shafranov VD 1957 About stable magneto-hydrodynamic configurations *J. Exp. Theor. Phys*. 33(3(9)) pp 710-722 (in Russian)
[39] Anderson O, Baker WR, Bratenahl A, Furth HP, Kunkel WB 1959 Hydromagnetic capacitor *J. Appl. Phys* 30(2) pp 188-196
[40] Artzimovich LA, Lukianov SYu 1978 *Motion of Charged Particles in Electric and Magnetic Fields* Moscow: Nauka (in Russian)
[41] Landau LD, Lifshits EM 1961 *Theory of Field* Moscow: Nauka (in Russian)
[42] Barry JD 1980 *Ball Lightning and Bead Lightning* New York and London: Plenum Press
[43] Uman MA 1969 *Lightning* New York: McGraw-Hill
[44] Nikitin AI, Nikitina TF, Velichko AM 2010 Corona discharge and levitation of ball lightning *Electrichestvo* 3 pp 16-22 (in Russian)
[45] Nikitin AI, Leipunsky IO, Nikitina TF 2010 A role of eddy currents in creating of lifting force by an autonomous emitter of radio waves *Prikladnaya Fizika* 2 pp 15-22 (in Russian)
[46] Nikitin AI, Velichko AM, Nikitina TF, Stepanov AI, Stepanov IG 2016 Observation and analysis of ball lightning group flights *Proc. 5th International Conf. "Atmosphere, Ionosphere, Safety" (AIS - 2016)* Kaliningrad pp 274-278
[47] Ternov IM, Mikhailin VV, Khalilov VR 1980 *Synchrotron Radiation and its Applications* Moscow: MGU Publisher (in Russian)
[48] Nikitin AI 2000 Creation of ball lightning at developing of linear lightning *Electrichestvo* 3 pp 16-23 (in Russian)
[49] Bychkov V 2010 Artificial and natural fireballs as combustion objects *IEEE Transactions on Plasma Science* 38(12) pp 3289 – 3290
[50] Emelin SE, Semenov VS, Bychkov VL et.al. 1997 Some objects formed in the interaction of electrical discharges with metals and polymers *Tech. Phys* 42(3) pp 269-277
[51] Emelin S, Bychkov V, Astafiev A, Kovshik A, Pirozersky A 2012 Plasma combustion nature of artificial ball lightning *IEEE Transactions on Plasma Science* 40(12) pp 3162-3165
[52] Bychkov VL, Chernikov VA, Osokin AA, Stepanov AI, Stepanov IG 2015 Modeling of artificial ball lightning with a help of capillary discharge *IEEE Transactions on Plasma Science* 43(12) pp 4043-4047
[53] Radzig AA, Smirnov BM 1980 *Reference book on atomic and molecular physics* Moscow: Atomizdat publishers (in Russian)
[54] Smirnov BM 1987 *Introduction to plasma physics* Moscow: Nauka publishers. (in Russian)
[55] Bychkov VL, Bychkov AV, Timofeev IB 2004 Experimental production of long-lived luminous organic polymeric objects *Technical Physics* 49(1) pp 128-133
[56] Bychkov VL, Bychkov AV, Timofeev IB 2006 Experimental modeling of long-lived luminous formations in air on a basis of polymeric organic materials *Khimicheskaya Fizika* 25(4) pp 89-96 (in Russian)
[57] Amirov AKh, Bychkov VL 1994 ANOVA of Several Parameters of the SKB Data Bank on Ball Lightning *Physica Scripta* 50 pp 93-96
[58] Amirov AKh, Bychkov VL, Bobkov SE 1998 On the dependence lifetime- diameter for ball lightnings *Physica Scripta* 58 pp 56-60
[59] Kolosovskii OA 1981 Observation of ball lightning track on the window glass *Zh. Tekh. Fiz*. 51 pp 856-858 (in Russian)
[60] Bychkov VL, Nikitin AI, Ivanenko IP, Nikitina TF, Velichko AM, Noisikov IA 2016 Ball lightning passage through a glass without breaking it *J. of Atmospheric and Solar-Terrestrial Physics* 150 pp 69-76

On modeling of "plasmoid" created by electric discharge

V L Bychkov[1], S V Anpilov[2], N P Savenkova[3], V Stelmashuk[4], P Hoffer[4]

[1]Leading researcher, Lomonosov MSU, Faculty of Physics, Moscow, Russia
[2]Junior researcher, Lomonosov MSU, Faculty of Computational Mathematics and Cybernetics, Moscow, Russia
[3]Leading researcher, Lomonosov MSU, Faculty of Computational Mathematics and Cybernetics, Moscow, Russia
[4]Institute of Plasma Physics, Za Slovankou 3, 182 00 Prague 8, Czech Republic

E-mail: stelmashuk@ipp.cas.cz, bychvl@gmail.com

Abstract. Gas dynamic modeling approach was applied to "plasmoid" formed in experimental electrical discharge breakdown in a cathode ceramic tube filled by a conductive liquid. Measured parameters were applied to the model in order to improve it. Qualitative agreement observed between behavior of the experimental "plasmoid" and model results. Temperature distribution results also agree with experiment in which temperature of 2000 K was registered.

1. Introduction

Numerous experiments performed in recent years [1-4] have led to observations of autonomous luminescent objects (so-called "plasmoids") being formed in electrical discharge breakdown in conductive liquid. The current article research work is devoted to modeling of the new experiment. Principle scheme of the experimental setup is shown in Fig. 1.

Electric discharge between two electrodes generates approximately 11.67 kJ of energy (see Fig. 2 for measured electric current and voltage). In described setup one electrode is placed on the bottom of a vessel filled with conductive liquid (water + NaBr with the conductivity of 3 mS) while the second electrode is placed over the liquid surface. Resulting generated energy is distributed between molecular excitation of the liquid thus forming active molecules and vapor – and formation of luminescent layer of plasma on the liquid surface around the electrode.

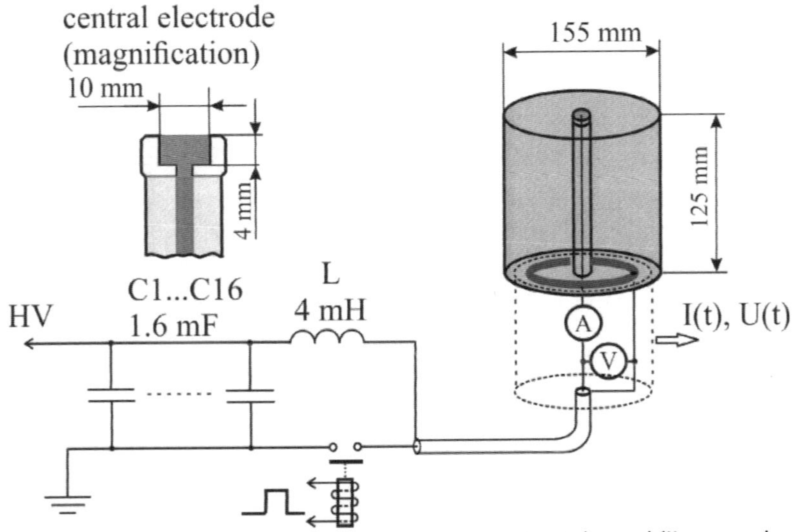

Figure 1. Scheme of the experimental setup for the "plasmoid" generation.
Central electrode - with the ceramic tube.

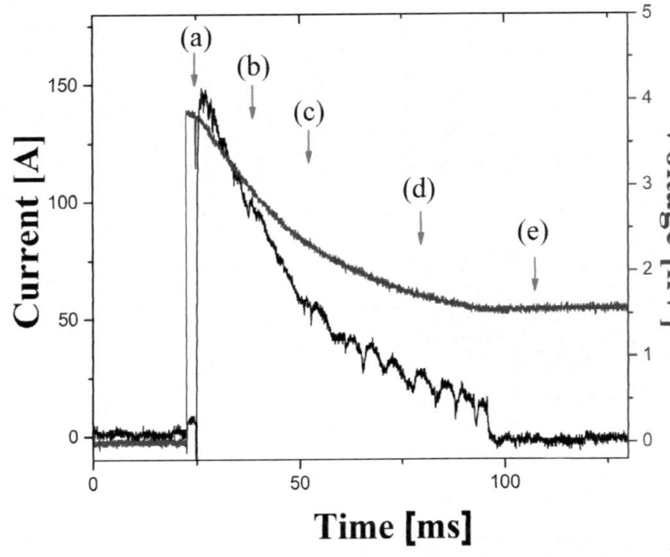

Figure 2. Evolution of voltage and electric current between electrodes.

Detailed examination of photos of discharge processes show that net-like glowing sparks connecting the liquid surface of the vessel with the liquid surface of the cathode appears. All the liquid in the ceramic tube is blown up to the upper region adjacent to the cathode transforming into the "foam" like mix of water bubbles and vapor. Jet formation emitting a plasma along its axis is observed afterwards. Then the formation reaches shape of somewhat like a glowing sphere which finally evolves into a vortex toroid.

In Fig.3 one can see evolution in time of our "plasmoid", and in Fig.4 schematics of experimental media motion is presented.

"Gatchina" discharge experiments [2] reveal complicated chemical processes inside the "plasmoid" resulting in temperature rise up to 2000 K. Explanation was given to long life of plasmoids by Egorov

IOP Conf. Series: Journal of Physics: Conf. Series **996** (2018) 012012 doi:10.1088/1742-6596/996/1/012012

and Stepanov in [3] based on formation of stable clusters consisting of two hydrated ions of opposite signs and water molecules preventing hydrate ions recombination and this extending the life-time of ions existence. In reference [4], published after our [5] was shown that the plasmoid is transformed in the ring during evolution. This corresponds to results of gas-dynamic calculations [5-6]. In them was proposed that such objects have a structure of toroidal vortexes formed by a jet pulse which are known to be stable and long-lived [1].

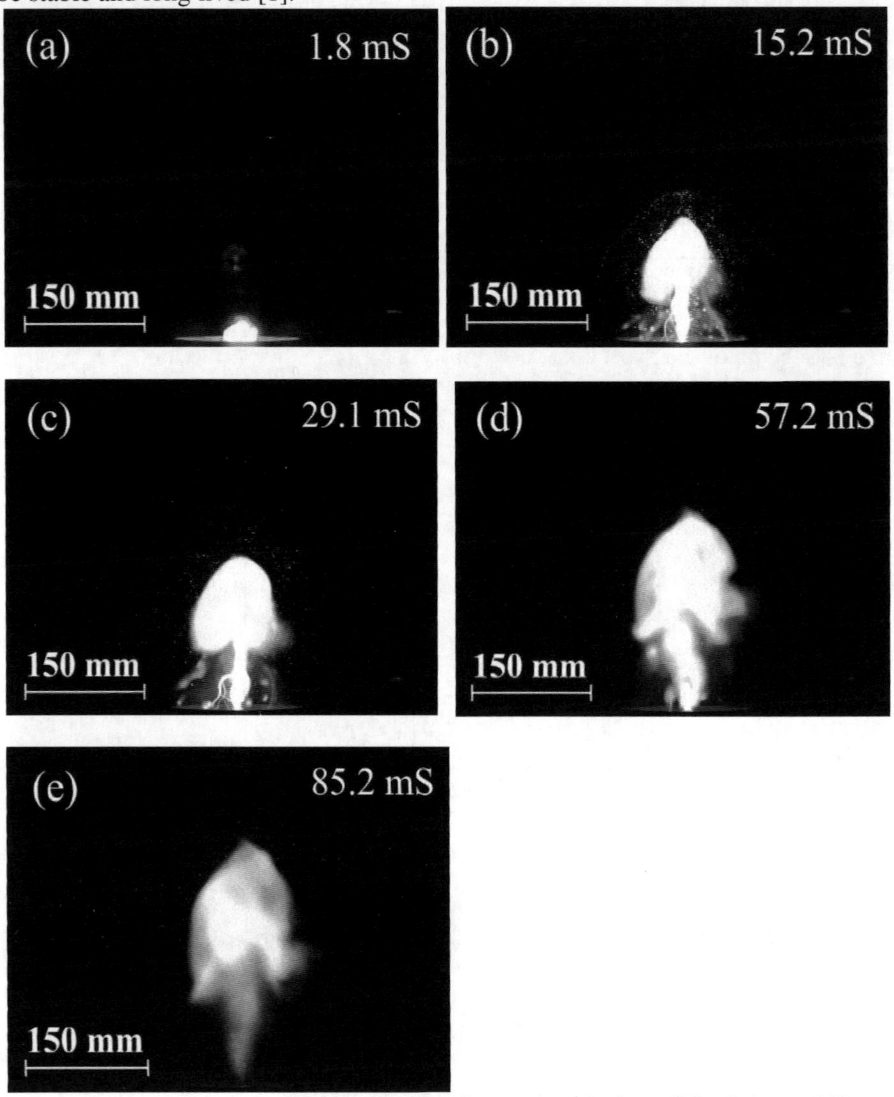

Figure 3. Experimental evolution of the "plasmoid"

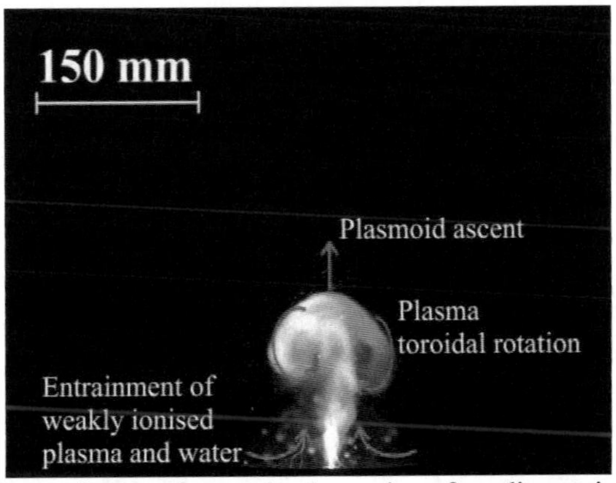

Figure 4. Experimental schematics of media motion.

2. Mathematical model

Presented model is based on the model previously used by authors [5-6] which is based on unsteady viscous gas dynamic approach. Partial differential equations (1) – (3) were solved using Finite Volume Method [7].

Generated heat was applied as a source in region around central electrode; mass inflow was applied at ceramic tube circular surface area so that total gas media mass would be the same as the mass of water contained in the tube.

We will consider therefore Navier-Stokes equations with respect to axis-symmetry in a form:

$$\frac{\partial \mathbf{U}}{\partial t}+\frac{\partial \mathbf{A}}{\partial r}+\frac{\partial \mathbf{B}}{\partial z}=\mathbf{C}, \tag{1}$$

where

$$\mathbf{U}=\begin{bmatrix}\rho\\ \rho v_r\\ \rho v_z\\ \rho h\end{bmatrix},\ \mathbf{A}=\begin{bmatrix}\rho v_r\\ p+\rho v_r^2\\ \rho v_r v_z\\ \rho v_r h\end{bmatrix},\ \mathbf{B}=\begin{bmatrix}\rho v_z\\ \rho v_z v_r\\ p+\rho v_z^2\\ \rho v_z h\end{bmatrix},\ C=\begin{bmatrix}-\rho v_r / r\\ -\rho v_r^2 / r+\nu\left(\partial^2 v_r / \partial r^2+\partial^2 v_r / \partial z^2\right)\\ -\rho v_r v_z / r+\nu\left(\partial^2 v_z / \partial r^2+\partial^2 v_z / \partial z^2\right)\\ 0\end{bmatrix}; \tag{2}$$

$$p=\frac{\rho}{\mu}RT. \tag{3}$$

Here (v_r, v_z) - velocity components, ν - viscosity, ρ - density, p - pressure, h - enthalpy. These equations will be solved in the cylindrical area 50 cm in radius and 100 cm in height, inlet radius of 5 mm corresponds to ceramic pipe surface radius.

Duration of the «jet impulse» Δt was set to 110 mc according to experimental measures. Boundary conditions are described in the Fig. 5.

Unsteady heat generation was set as:

$$q_v=\begin{cases}\dfrac{1.167\cdot 10^4}{\Delta t}\ \text{W},\ 0\le t\le \Delta t\\ 0,\ t>\Delta t\end{cases}$$

Formation of gas media out of water in ceramic pipe was set as mass flow:

$$m = \begin{cases} \dfrac{3.14 \cdot 10^{-4}}{\Delta t} \text{ kg}, \ 0 \le t \le \Delta t \\ 0, \ t > \Delta t \end{cases}$$

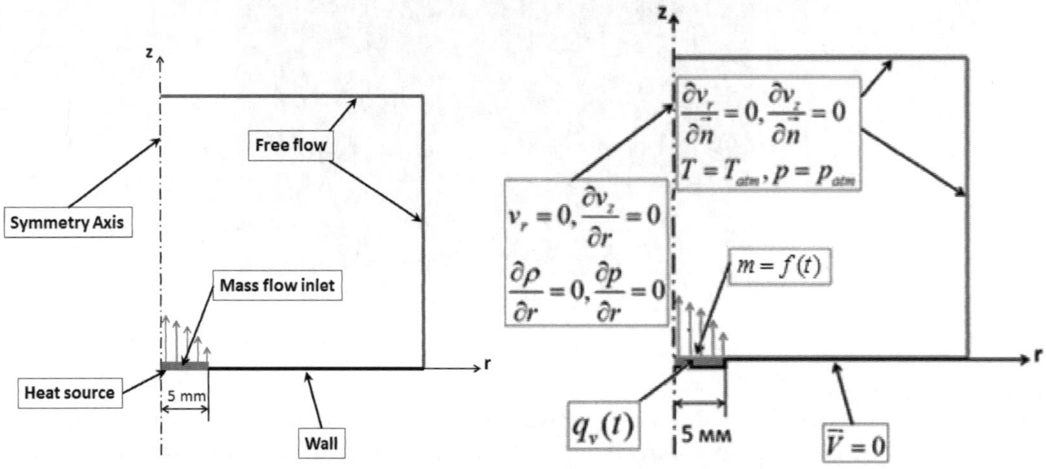

Figure 5. Task definition: a – boundaries, б – boundary condition.

Initial conditions correspond to still air at standard conditions.

3. Simulation results

Our modeling results at different moments of time are presented in Fig. 6-7. The fact that formation of heated volume can be seen with vortex structure in its upper part (Fig. 7) agrees with conclusions in work [5] – also vortex diameter is very close to the one observed in the experiment. Vortex toroid structure lifetime is estimated about 0.25 s which is close to measured times of this experiment and [2]. One must admit the remaining hot "leg" in the bottom – which brings us to conclusion that discharge heating in the whole conducting gas volume should be considered in further investigations.

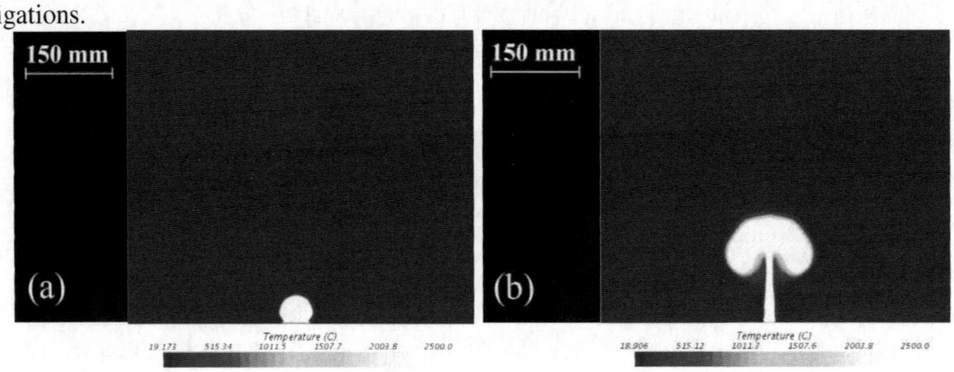

IOP Conf. Series: Journal of Physics: Conf. Series **996** (2018) 012012 doi:10.1088/1742-6596/996/1/012012

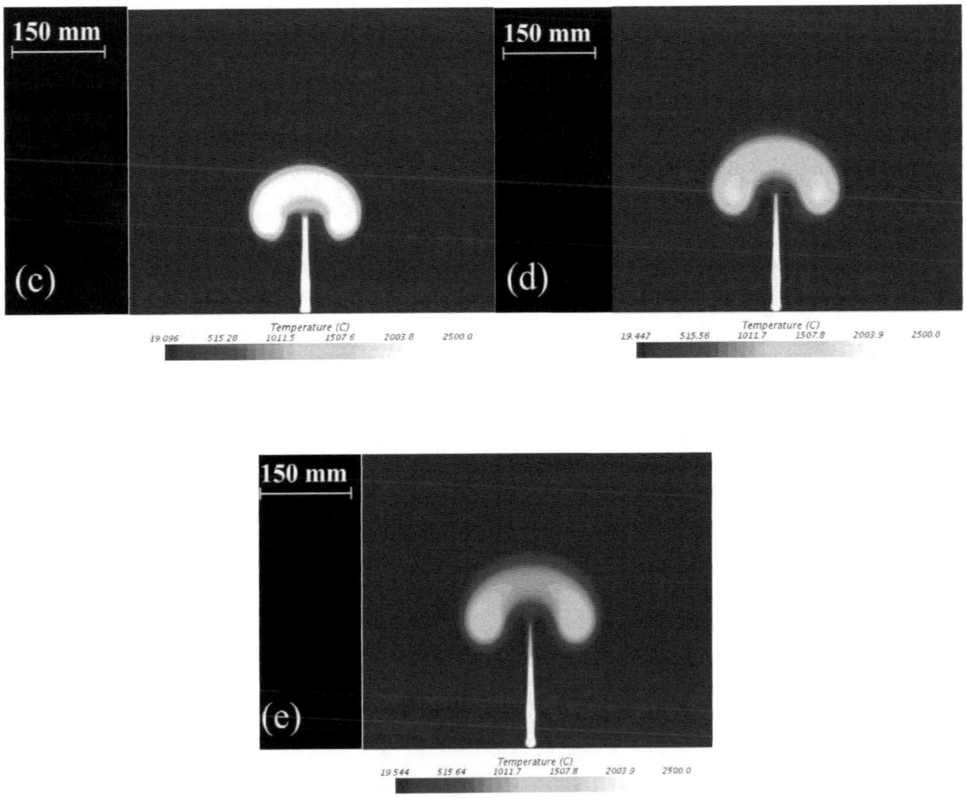

Figure 6. Calculations. Temperature distribution at time moment (a) 33.418 ms; (b) 78.131 ms; (c) 110.987 ms; (d) 154.138 ms; (e) 170.938 ms.

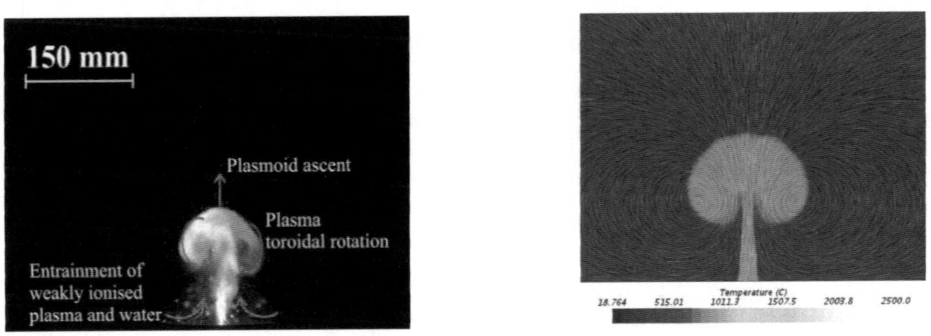

a) *b*)

Figure 7. *a*– Schematics of media motion a –experimental, *b* – direction of media motion obtained in calculations.

4. Conclusions

Results of modeling of discharge "plasmoids" observed in presented experiment were presented in the current paper. Numerical experiments were carried out for proposed model of gas heated around the central electrode.

Simulation results even under carried out simplifications are in agreement with "plasmoid" dynamic observed in the experiment. Temperature distribution also agrees with experiment [2] in which temperature of 2000 K was measured.

Existing disagreement brings us to conclusion that the model should be updated with more complicated heat release in the whole volume of the "plasmoid" with respect to plasma radiation effects.

References

[1] Shabanov G D, Zherebtsov O M, Sokolovskyi B Yu *Khimicheskaya Fizika*, 2006. **25** (4), P.88.

[2] Egorov A I, Stepanov S I *Tech. Phys.* 2002. **47(12)**, P.1586.

[3] Juettner B, Noak S, Versteegh A, Fussmann G *Proc. 28th Int. Conf. on Phenom. in Ioniz. Gases*. Prague, Czech Republic, 2007. P.2234.

[4] Stepanov S I, Ultrasonic probing of plasmoid. *J. Tekhnicheskoi Fiziki*, 2014. **84(1)**, P.114.

[5] Bychkov V L, Anpilov S V, Savenkova N P, *Russian Journal of Physical Chemistry*, 2014. **8(1)**, P.55.

[6] Bychkov V L, Savenkova N P, Anpilov S V and Troshiev Y V, *IEEE Transactions on Plasma Science*, 2012. **40(12)**, P.4.

[7] Chung T J, *Computational Fluid Dynamics*. Cambridge University Press 2002.

IOP Conf. Series: Journal of Physics: Conf. Series **996** (2018) 012013 doi:10.1088/1742-6596/996/1/012013

On electro-hydrodynamic effects over liquids under influence of corona discharge

V L Bychkov[1], V I Abakumov[2], A R Bikmukhametova[2], V A Chernikov[3], D A Safronenkov[4]

[1]Leading researcher, Lomonosov MSU, Physical faculty, Moscow, Russia
[2]Graduate student, Lomonosov MSU, Physical faculty, Moscow, Russia
[3]Associated professor, Lomonosov MSU, Physical faculty, Moscow, Russia
[4]Student, Lomonosov MSU, Physical faculty, Moscow, Russia

E-mail: bychvl@gmail.com

Abstract. Electrohydrodynamic effects over liquids under high voltage electrode are considered in experiments with corona discharge. Simple theory is applied for description of a funnel appearance over a liquid is presented. New types of electrohydrodynamic instabilities are revealed.

1. Introduction

Present investigations are devoted to formulation and undertaking of experiments with corona discharges realized over surfaces of different liquids. Investigations when a surface of one of the electrodes is covered by a liquid or is an electrode itself are of interest for hydrocarbon fuel activation and search of elimination of undesirable liquids in engines, problems of ecology and disinfection with a help of electric discharge. Such investigations are developing now, that is a reason for undertaking of these investigations. In this work we concentrated on electric-hydrodynamic features of corona discharges which are practically uninvestigated yet. However, there is limited information about any discharges over surfaces of liquids [1,2] and it is obtained mainly for electrolytes or water. From the application point of view it is interesting to know how a surface of a liquid fuel acts under an impact of gas discharges. Therefore our work is devoted to investigations of electric-hydrodynamic (EHD) effects caused by corona discharges over liquid. We made several series of experiments with positive and negative coronas over surfaces of tap water, distilled water, alcohol, glycerin, kerosene, and their mixtures.

2. Electro-hydrodynamics phenomena with corona discharges

Corona discharges are typically obtained by applying a high potential between a small diameter wire, charged negatively or positively and plate or coaxial cylinder [3] charged appositively. Investigations of these discharges and their applications continue for more than hundred years. Recently corona discharges began to be applied for problems of plasma aerodynamics [2].

The principle scheme of the experimental device is represented in Fig.1.

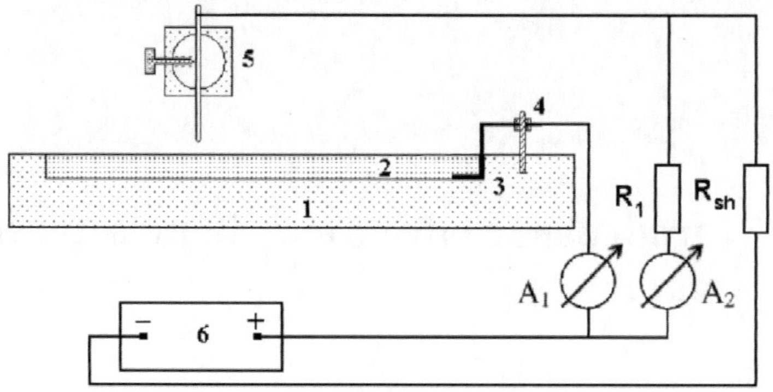

Figure 1. Principle scheme of experimental device.
1-Ditch, 2- Liquid (water, alcohol, kerosene), 3-Anode, 4- Post, 5- Cathode, 6-Power supply

It consists of a ditch filled with a liquid (water, alcohol, kerosene, etc), and electrical circuit. Upper electrode or a set of electrodes of diameter 0.9 mm (with radius of the tip-0.2 mm) or 2 mm (with radius of the tip-0.4 mm) was located at height 5-15 mm over a liquid surface. A distance between electrodes in upper multi-electrode formulation was 11 mm. Electrodes were under positive or negative potentials. Ditches were either metallic or dielectric. The metallic ditches were: cylindrical ∅ 130 mm, height 18 mm, rectangular 37×70×122 mm; the dielectric plastic ditch was: rectangular 45×95×130 mm.

A current in a discharge was measured by a milliampere meter A1, a voltage was measured by a chain consisting of resistance R1 and a milliampere meter A2. We measured Ampere and Volt characteristics of the discharge and represented results in A-V2 coordinates typical for corona discharges [2].

Experiments have been carried out with a help of different cavities: rectangular – dielectric, cylindrical – dielectric and steel. Sizes of the rectangular cavity, were 70×20×15 mm (length, width and depth). Characteristics of cylindrical cavities were ∅125, 10 mm and ∅185, 20 mm, respectively. An investigating liquid was poured into a cavity, a negative or positive electrode was located directly over the liquid, a distance between the electrode and the liquid could be varied in a range (1- 30) mm. The upper electrode in the case of rectangular cavity was placed at the angle of 50° to the vertical. Its diameter was 3 mm, its angle at the vertex was 30° and a radius of a tip was 0,4 mm. An electrode of another polarity was located directly in the liquid. (A cavity was used as an electrode itself in the case of metallic cavity). It was connected with a feeding source through a hermetic unit. In this case the liquid becomes the second electrode. A high voltage generator was used as the feeding source, it allowed to measure a voltage on the electrodes from 2 kV to 50 kV with a step of 250 V. Typical range of the voltage was 5-25 kV. Appearance of the spark discharge was accompanied with a drop of the voltage at the capacity, which was detected by the voltmeter. In this device we used a ballast resistance, which value was 100 MΩ. Typical values of a current were 50-100μA.

3. Experiments with discharges over tap water surface

Our experiments have shown that application of negative and positive corona discharges over water surfaces in plastic or metallic cavities lead to appearance of a rotating funnel, it was evidently caused by the ion wind, see Fig.4.2-4.3. Vortices over surfaces of liquids appeared in case of non-central position of the upper electrode with respect to the cavity and in rectangular cavities.

Figure 2. Appearance of a funnel over water surface covered with wooden powder, metallic cavity, a distance between the upper electrode and a surface 5 mm, electrode central

Figure 2 represent appearance of a funnel over water surface covered with wooden powder. Figures 3-4 represent a transformation of a funnel into water surface excited by surface waves at rise of the voltage.

Figure 3. Appearance of a funnel over a water surface, plastic cavity, a distance between the upper electrode and a surface 5 mm. Negative corona.

Figure 4. Appearance of waves over a water surface, plastic cavity, a distance between the upper electrode and a surface 15 mm. Negative corona.

4. Experiments with alcohol

In case of the alcohol in plastic or metallic cavities corona discharges lead to appearances of different EHD effects over the surface of the liquid.

Funnels appeared at both polarities of the upper electrode. In Fig.5. we represent such a phenomenon.

Figure 5. Appearance of a cavity over alcohol surface, metallic cavity, a distance between the upper electrode and a surface 5 mm. Positive corona.

In case of the negative upper electrode firstly a rotating funnel usually appeared (it was evidently caused by then ion wind). Later some nuclei of jets in a form of tips were observed. Then bursting jets arose from the nuclei. The sprouts develop on sides of the funnel. This finishes either appearance of jets, see Figures 6-9, or liquid columns, see Figure 10-11. Usually this takes place at high applied voltage (20-30 kV). These columns are either standing on a place of the post funnel center or make some rotation motion being on a side of a funnel.

Liquid columns have sometimes widening at tip, sometimes there appears a violet-bleu glow on the column tip, see Figures 10-11. It speaks about breakdown field conditions realization near the tips.

Images were obtained with a help of digital video camera Sony DCR-TRV730 with frame duration 33 ms. Appearance of jets and columns stopped at the upper electrode distance of 17-18 mm from the surface.

Similar phenomena are known in EHD from literature [46]. They are connected with a development of Rayleigh, Taylor, Tonks and Frenkel electrostatic instabilities on surfaces of charged drops and jets when Coulomb forces of accumulated charges become greater than the surface tension.

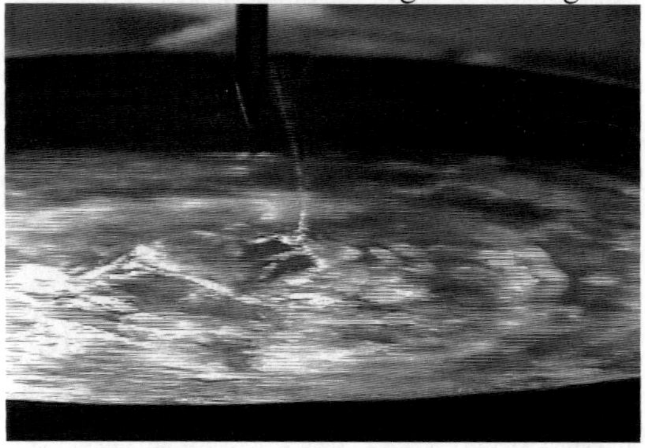

Figure 6. Appearance of a jet over a double distilled water surface, plastic cavity, a distance between the upper electrode and a surface 5 mm. Negative corona.

Figure 7. Appearance of a jet over alcohol surface, metallic cavity, a distance between the upper electrode and a surface 7 mm. Negative corona. First stage.

Figure 8. Appearance of a jet over alcohol surface, metallic cavity, a distance between the upper electrode and a surface 7 mm. Negative corona. Second stage.

Figure 9. Appearance of a jet over alcohol surface, metallic cavity, a distance between the upper electrode and a surface 7 mm. Negative corona. Third stage.

Figure 10. Appearance of a column over alcohol surface, metallic cavity, a distance between the upper electrode and a surface 7 mm. Negative corona.

Figure 11. Appearance of a discharge between the upper electrode and the column over alcohol surface, metallic cavity, a distance between the upper electrode and a surface 7 mm. Negative corona.

5. Short theoretical discussion of obtained results

Let us consider impact of the ion wind to a surface of a liquid, and suppose that the ion wind creates a funnel. A scheme of the corona discharge and a liquid are represented in Fig.4.27.

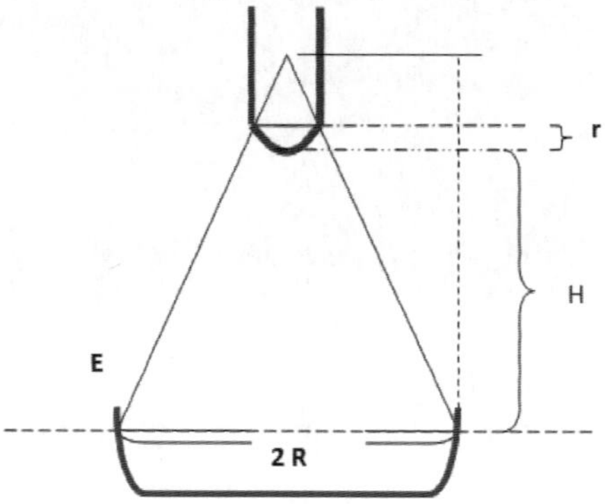

Figure 12. A scheme of the ion wind impact on a liquid surface

Appearance of the funnel can be a result of the ion wind impact on the liquid surface. Disregarding air heating by the corona discharge one obtains from the Euler equation a following equation for air stream velocity [5,6]

IOP Conf. Series: Journal of Physics: Conf. Series **996** (2018) 012013 doi:10.1088/1742-6596/996/1/012013

$$\frac{d}{dx}\left(\frac{1}{2}\rho u^2\right) = eE\,n_i = \frac{I}{\mu_{ion}A(x)}, \tag{1}$$

here ρ is air density, e- electron charge, E - electric field, μion-a mobility of ions, I –discharge current, A(x) – a cross section of air stream at a distance x from the tip of the upper electrode. Taking an area of the ion wind as a cone with a radius of a base Rs, and the cone side surface touches a base of the tip's hemisphere with the radius r0 one can obtain a following equation for a pressure head, see Figure 12, one gets:

$$\frac{\rho u^2}{2} \approx \frac{I \cdot d}{\mu_{ion}\pi \cdot r_0 \cdot R_s}, \tag{2}$$

where d is a distance between a surface of liquid and a tip of the upper electrode. For experimental conditions $I \approx 5\cdot 10^{-5}$ A, $d \approx 10$ mm, $r_0 \approx 0{,}2$ mm, $R_s \approx 3$ mm, $\mu_{ion}=2\cdot 10^{-4}$ $m^2\ V^{-1}s^{-1}$ one obtains from (3): $\frac{\rho\, u^2}{2} \approx 1{,}3\cdot 103$ Pa.

Using a scheme in Figure 12 one obtains the following equation

$$\delta\, p = 1/2\cdot \rho u^2 = I\cdot (R-r)^2/(\pi\cdot \mu_{ion}\cdot H\cdot r\cdot R). \tag{3}$$

Inserting air density $\rho=1{,}2$ kg/m³, experimental values of the current I=50 μA, the radius of the funnel R=2 mm, the radius of the tip r =0,4 mm of the upper electrode, the distance between the tip and the surface H= 7 mm and air ion mobility $\mu ion=2\cdot 10^{-4}$ $m^2V^{-1}s^{-1}$ into (3) one obtains the following value of an average velocity in conditions of our experiment u≈ 7.8 m/s. It is in a reasonable agreement with measurements [7-8] where the values of the ion wind at farer distances from the upper tip were about 3-5 m/s.

The pressure drop given by (3) is $\delta\, p \approx 36$ Pa.

From another point of view one can find it using a volume of the compressed liquid at a formation of the funnel:

$$\delta\, p = \rho\cdot g\cdot h\cdot S/S = \rho\cdot g\cdot h$$

where h is a depth of the funnel. In case of water in our experiment h ≈ 3 mm, and $\delta\, p \approx 30$ Pa, it is in reasonable agreement with the estimate given on a basis of the ion wind. So one can conclude that the funnel is created by the ion wind.

According to [1,3] an electric field E_{max} at a distance x between the tip of the parabolic form with curvature radius r being and the perpendicular plane at a distance d is connected with the voltage V between the tip and a plane by the formulae

$$E_{max} = 2V/((2x+r)\ln(2d/r+1)). \tag{4}$$

Electric field in the corona discharge can be described by this formula [3] when there is no developed discharge channel.

One can see that the electric field influences the charged surface of the liquid. It will attract charged surface, against the gravitational force and the force of surface tension, causing the electric instability. It is so called Tonks – Frenkel instability of the charged surface of a liquid [4,6]. A condition for realization of this instability has a form

$$E^2 > 2\sqrt{\rho\cdot g\cdot \alpha}/(\varepsilon\varepsilon_0). \tag{5}$$

E is electric field strength near a surface of a liquid, ε 0 is air dielectric permeability, α is a surface tension of the liquid, ρ is its density. Using this formula one can obtain thresholds of Tonks-Frenkel instability over different liquids, such estimates for the critical electric fields are represented in Table 1.

Table 1. Values of the critical electric field for Tonks-Frenkel instability

Liquid	Distilled water	Alcohol	Glycerin	Kerosene
E, MV/ m	2.4	1.7	2.5	2.1

Looking at the Table one can see that the thresholds for glycerin and distilled water are close to those of the air breakdown, which in real conditions varies in the range E≈2.6-3/0 MV/m. At the same time its threshold in alcohol is E≈1.7 MV/m -much smaller. This shows that the development of air breakdown and formation of plasma channel can damp the development of the instability. However, one can expect a development of this instability in alcohol and kerosene and their mixtures. What we see in our experiments.

Using formula (4) and Figure 12 one can understand why jets and columns can be easier originated at the edge of the funnel. Observations show that the edge height reaches 1 mm size. From (4) one obtains a ratio of electric fields between the tip and the funnel bottom and the tip and the edge:

$$E_{edge} / E_{bottom} \approx (1+(\Delta s_1 + \Delta s_2)/d),$$

where Δs_1 is the depth of the funnel and Δs_2 is the size of the edge. If $\Delta s_1 \approx 1.5$ mm, $\Delta s_2 \approx 1{,}0$ mm and d=7 mm then $E_{edge} / E_{bottom} \approx 1.35$, so conditions for the development of the instability at the edge of the funnel are favorable. That we also see in the experiment.

Analysis of Volt 2 - µAmpere characteristic of obtained experimental data shows that appearance of jets and columns over alcohol in negative corona take place at voltages when in the positive corona the discharge channel is already realized. So, conditions in the negative corona are similar to those realizing in the capacitor [3,5] with high electric field causing the instability and in the positive corona – in the quasi-neutral plasmas with lower electric field, which cannot cause a development of this instability.

Appearance of bursting jets and destruction of columns can be connected with development of instabilities of charged drops (we consider that tips of these structures represent quasi-drops). Conditions of their development have a form [4]:

$$\varepsilon_{air}\varepsilon_0 E^2 / 2 \geq 0.7 \cdot \alpha / R$$

in the case of large deformations (Rayleigh problem) and

$$\varepsilon_{air}\varepsilon_0 E^2 / 2 \geq 2 \cdot \alpha / R$$

in the case of small deformations, E is electric field strength on the drop surface, εair is the dielectric constant of air, α is the surface tension coefficient, R is the radius of the drop. Assuming that the alcohol drop radius is equal to the radius of the column one obtains E ≥2.6÷4.5·MV/m. These values are close to the typical breakdown values realized in air discharge near the electrodes. So we can connect these destruction mechanisms with the phenomena observed in the discharge.

Appearance of jets and columns over a surface in negative corona discharge case can be connected with a time of electric charge relaxation in a liquid. According to electrodynamics this time is $\tau \sim \varepsilon_{liq}\varepsilon_0 / \sigma_{liq}$, here ε_{liq}, ε_0 are a dielectric constant of a liquid and vacuum, respectively, σ liq is electric conductivity of a liquid. For example, inserting values of corresponding values one obtains a typical time of charge relaxation in alcohol τ~22 s and τ~7·10-7 s in tap water. So in case of alcohol a state of a quasi-capacitor can be realized, and a surface of liquid can stay charged during a long time. However, ions in the positive corona are transported to the lower electrode quicker than in negative corona, and they will neutralize the charge in the liquid. So the situations with positive and negative coronas are different. It is well known that a liquid is drawn to a region of a high field in a capacitor that can be the reason for appearance of jets and columns in case of negative corona over a surface of alcohol.

6. Conclusions

Creation of corona discharges of positive and negative polarity has been realized over tap and distilled water, alcohol, glycerin and their mixtures.

Special devices for undertaking of experiments with several electrodes have been developed.

Conducted experiments and their analysis show that the corona discharge over different liquids can cause different hydrodynamic phenomena including appearance of the ion wind and different instabilities.

Appearance of funnels is a result of the ion wind this was proved on the basis of the Euler equation. Funnels appear both in positive and negative coronas and independent of the discharge type.

Appearance of liquid jets and columns is a result of the Tonks–Frenkel instability in conditions of negative and positive corona. This phenomenon depends on a value of electric field strength and is better realized in liquids with low density and surface tension.

Separation of these jets and columns in drops are connected with instabilities of charged drops.

Experiments show that conditions in the negative coronas in air are more favorable for creation of hydrodynamic phenomena than those in the positive coronas, since in the negative coronas can be achieved higher values of electric fields at which is still no realization of the discharge plasma channel connecting the electrodes.

These experiments helped to clarify modes of discharges realization and their connection with appearance of funnels on the surfaces of liquids and columns in case of alcohol.

References

[1] Lebedev Yu A, Plate N A, Fortov V E *Encyclopedia of low temperature plasma. Applied Plasma Chemistry*. Ed. 2009 **9(5)** Yahys-K publishers Moscow

[2] Bychkov V, Chernikov V, Ershov A Esakov I, Kostiuk A 2009 *Corona discharge over a surface of a liquid*. AIAA-2009-1554. 47th AIAA Aerospace Sciences Meeting. 5-8 January 2009. Orlando World Center Marriott. Orlando. Florida.

[3] Raizer Yu P *Gas discharge physics*. Nauka, Moscow 1992.

[4] Saranin V A *Equilibrium of liquids and its stability*. Institut Kompiuternykh Issledovanii Publishers. Moscow 2002.

[5] Landau L D, Lifshitz EM *Electrodynamics of continuous media*. Moscow, Nauka 1982.

[6] Levich V G *Physical and Chemical Hydrodynamics. Fizmalit Publishers*. Moscow 1959.

[7] Lacoste D A, Pai D, Laux CO 2004 AIAA-2004-354. 42-nd AIAA Aerospace Sciences Meeting and Exhibit. 5-8 January 2004, Reno, Nevada.

[8] Kozlov B A, Solov'ev VI *Zhurnal Tekhnicheskoi Fiziki* 2007 **77(7)** P.76.

IOP Conf. Series: Journal of Physics: Conf. Series **996** (2018) 012014 doi:10.1088/1742-6596/996/1/012014

On the physics of electron transfer (drift) in the substance: about the reason of "abnormal" fast transfer of electrons in the plasma of tokamak and at known Bohm's diffusion

I A Boriev

The Branch of Talrose Institute for Energy Problems of Chemical Physics of Russian Academy of Sciences, Chernogolovka, 142432, Russia

E-mail: boriev@binep.ac.ru

Abstract. An analysis of the problem of so-called "abnormal" fast transfer of electrons in tokamak plasma, which turned out much faster than the result of accepted calculation, is given. Such transfer of hot electrons leads to unexpectedly fast destruction of the inner tokamak wall with ejection of its matter in plasma volume, what violates a condition of plasma confinement for controlled thermonuclear fusion. It is shown, taking into account real physics of electron drift in the gas (plasma) and using the conservation law for momentum of electron transfer (drift), that the drift velocity of elastically scattered electrons should be significantly greater than that of accepted calculation. The reason is that the relaxation time of the momentum of electron transfer, to which the electron drift velocity is proportional, is significantly greater (from 16 up to 4 times) than the electron free path time. Therefore, generally accepted replacement of the relaxation time, which is unknown a priori, by the electron free path time, leads to significant (16 times for thermal electrons) underestimation of electron drift velocity (mobility). This result means, that transfer of elastically (and isotropically) scattered electrons in the gas phase should be so fast, and corresponds to multiplying coefficient (16), introduced by D. Bohm to explain the observed by him "abnormal" fast diffusion of electrons.

1. Introduction

It is known that transfer of electron beams in tokamak plasma occurs much faster (in ~10 times) than expected result of accepted calculation, and therefore it was called as "anomalous" transfer of electrons. Such fast transfer of electrons leads to unexpectedly rapid destruction of inner wall of tokamak chamber by hot electrons and to ejection of wall matter in tokamak plasma, what violates adjusted plasma properties desired for realization of controlled thermonuclear fusion. This circumstance created a complicated problem of wall protection from hot electron beams, for solution of which considerable expensive efforts are used. For successful realization of controlled fusion it is necessary to reveal the cause of "anomalous" electron transfer in order to establish possible methods to suppress this damage by fast transfer of hot electrons. It is important, particularly, for implementation of controlled thermonuclear fusion within effected in France well-known ITER (International Thermonuclear Experimental Reactor) project, which is based on tokamak for plasma confinement at controlled fusion.

Paper gives theoretical explanation of the problem of "anomalous" (unexpectedly fast) transfer of electrons with the use of earlier developed, statistically correct theory of electron transfer and heating in the gas under electric field force [1-3]. This theory, which results are well confirmed by known experimental date [4-6], takes into account for the first time real physics of forced transfer (drift) of electrons in the gas phase and is based on the fundamental conservation laws of momentum and energy. Using the law of momentum conservation for transfer impulse of drifting electrons it is shown that electron drift velocity (and mobility) in gas phase for elastically scattered electrons should be much faster (from 16 to 4 times) than the value of accepted calculation [7].

Really, the reason of the problem of "anomalous" fast electron transfer is rather simple. At calculation of electron drift velocity, the value of which is proportional to relaxation time of electron transfer impulse (according to the main concepts of physics), it is generally accepted to substitute this time, which is not known a priory, by electron free path time, which may be calculated using known cross sections of electron scattering. Developed theory [1-3] gives that relaxation time of electron transfer impulse for elastically (and isotropically because of comparatively very small electron mass) scattered electrons is significantly larger (from 16 to 4 times) than the electron free path time [7]. This result directly follows from accomplished analyses [5, 6] of specific features of the field dependence of electron drift velocity in dense rear gases which was accurately established in [4]. As clear, such substitution should lead to corresponding significant underestimation of the value of electron drift velocity. Another important result of the theory is that an appreciable decrease of the difference between these two times (from 16 to 4) occurs at heating of drifting electrons (above their thermal energy coursed by substance temperature) due to their forced transfer under the force of the strong electric field.

Such wrong substitution of relaxation time of electron transfer impulse by the time of electron free path is generally substantiated by isotropic scattering of all drifting electrons during one free path time. As supposed, at such scattering also an isotropy (and full dissipation) of electron transfer impulse takes place at every one free path time. However, this substantiation is not correct, since at that no difference is made between electron transfer impulses and their thermal (isotropic) impulses, which don't contribute to the directed transfer of drifting electrons. To explain electron "anomalous" transfer it is important at first to separate the electron transfer impulses from their thermal impulses, and then apply to the electron transfer impulse the law of momentum conservation. At that, namely the fact of isotropic scattering of drifting electrons (because of their small mass with respect to the mass of substance particles) gives that the relaxation time of electron transfer impulse is significantly larger (namely, from 16 to 4 times) than the time of electron free path [7]. It is worth to mark, that sometimes it is assumed that the relaxation of electron transfer impulse may takes more than one possibly two or three) times of electron free path. Nevertheless, at calculation of electron drift velocity only one time of electron free path is always used instead of relaxation time of electron transfer impulse.

Therefore, obtained theoretic result means that the so-called "anomalously" fast electron transfer in tokamaks (and at known Bohm's diffusion) is a consequence of significant underestimation of electron drift velocity at accepted calculation, and observed fast electron transfer really must be so unexpectedly fast. Such result of simple statistically correct theory is well confirmed by theoretical analysis [5, 6] of known experimental data for electron drift velocity in dense rare gases [4]. This result, obtained due to taking into account the appearance of statistically average value of transfer impulse of drifting electrons and due to the use of momentum conservation law for this transfer impulse, shows that for correct explanation of observed physical processes it is always necessary take into account their real physics and use for their explanation known fundamental laws of physics. Known three conservation laws of classical physics and principles of quantum mechanics with the value of Planck's constant, as was recently shown [9], have their materialistic substantiation, since all they are determined by observed features of cosmic microwave background radiation produced by dark matter thermal seesaw motion at temperature 2.7K.

2. Real physics of electron drift in a gas under a constant weak and strong electric field

Understanding of the physics of electron transfer (drift) in the gas substance (including plasma) under the action of an electric field is important for the correct calculation of the electron drift velocity (V_d),

IOP Conf. Series: Journal of Physics: Conf. Series **996** (2018) 012014 doi:10.1088/1742-6596/996/1/012014

which is experimentally measured and which reveals real properties of the process of electron transfer. As clear, calculation of V_d dependence on the electric field should be based on the conservation law for the electron transfer impulse $P_E=mV_d$, where m is the mass of the electron, since namely this impulse is formed under the action of the electric field strength (E) [1-3]. Taking into account this basic physical property of the process of electron transfer (drift) in the substance is necessary for correct calculation of $V_d\,(E)$ since by comparing the result of this calculation with the experimental date for $V_d\,(E)$ dependence it is possible to reveal real properties of the electron transfer (drift) in the substance. The application of this approach to the calculation of $V_d\,(E)$ for the case of dense gas of rear atoms [5, 6] is given in the paper, what let revealed the physical reason of the so-called "anomalous" fast electron transfer, which is observed in the tokamak plasma and at well known Bohm’s diffusion.

Thus, the correct consideration of the physics of the process of electron transfer in a gaseous medium allows one to establish a quantitative relationship between the relaxation time (r_p) of the electron transfer pulse $P_E = mV_d$ with the time of their free path (τ): $\tau_p=Z\tau$ [5, 6, 7], where number Z must be larger than 1. It should be emphasized that the quantitative relationship between τ_p and τ is logical, since the relaxation (dissipation) of P_E occurs only at the scattering of electrons in the substance with frequency, which is determined by the value of τ.

This numerical connection (between τ_p and τ) has a fundamental importance for the correct calculation of V_d. The value of V_d, according to the conservation law for the transfer impulse $P_E=mV_d$, is proportional to τ_p, namely, $V_d=eE\tau_p/m$, where e is the electron charge, [1-3]. Indeed, since the rate of growth of the impulse P_E is equal to the force acting on the electron eE, and the rate of dissipation (relaxation) of P_E is generally determined through its relaxation time as P_E/τ_p, then the equality of these velocities at the stationary drift of electrons gives this simple expression for V_d value: $V_d=eE\tau_p/m$. However, the value of τ_p is not known a priori, because it is determined by different processes of dissipation of momentum and energy of the electron during its scattering, but the value of τ can be usually calculated from known scattering cross sections of electron. Historically (from the works of Drude and Lorentz at the end of the 19th century) it is simply supposed that the relaxation time of P_E may be replaced by the value of the electron free path time. Thus it is assumed that $\tau_p=\tau$, that is $Z=1$.

However, as can be shown by applying the conservation law to the transfer impulse $P_E=mV_d$, the number $Z=\tau_p/\tau$ for elastically (and isotropically) scattered electrons varies from 16 to 4, and the decrease of the Z value in this interval occurs with the increasing of the energy of the forced heating of the drifting electrons with increasing of E [3]. So, for correct calculation of V_d on the basis of simply determined value of τ, it is necessary establish the true value of Z. For calculation of Z value, it should be taken into account that the stationary electron transfer impulse is $P_E=mV_d=eE\tau_p=eEZ\tau$, and the increase of P_E in the direction of the field force action for each time τ is $eE\tau$. It is clear that in order to ensure the stationarity of electron transfer, it is necessary that for the time τ, during which all the drifting electrons are scattered isotropically, the same value of $eE\tau$ must be scattered against the action of the field strength, that is must be scattered Z^{-1} part of $P_E=eEZ\tau$. The correct calculation of the part of $P_E=eEZ\tau$ that is scattered against the action of the field, as presented in [7], gives that in the limit of a weak electric field, when the contribution of electron thermal velocities is significant, $Z=16$, and in the limit of a strong electric field, when the contribution of thermal velocities of electrons is negligible, $Z=4$. This theoretical result is fully confirmed by the known experimental data for the specific $V_d(E)$ dependence in dense inert gases, the most qualitative of which is given in the work done in the MEPhI [4]. The developed statistically correct ideas about the properties of forced transfer and heating of electrons in a gas [1-3] for the first time let to explain the specific effects observed for condensed inert gases: saturation (reaching a limiting value) of V_d in the limit of large E and a multiple increase of V_d at small additions of molecular impurities [5, 6].

Thus, revealed significant underestimation of generally calculated value of V_d (at assumption that $\tau_p=\tau$) let logically explain the so-called "anomalous" fast electron transfer, observed both in the tokamak plasma, and at known Bohm’s diffusion. By the way, D. Bohm, to explain observed unexpectedly rapid transverse transfer of thermal electrons (in crossed electric and magnetic fields), received the enlarging factor namely 16, which is known as the Bohm’s coefficient. It should be noted that such correction factor was obtained by D. Bohm by introducing an idea of chaotic motion of drifting electrons due to their magnetodynamics behavior. However, as clear, the use of this representation is unnecessary, because the isotropy of elastic scattering of drifting electrons (each time τ) really means their totally chaotic motion.

3. Statistically correct calculation of the value of electron drift velocity in the gas substance

Let's consider a drift of assembly of electrons (with mass m) under a constant electric field (E) in the gas of atoms (with mass $M>>m$). Observed (by electron current measurement) stationary drift velocity (V_d) of this electron assembly means an appearance (due to electric field force) of directed transfer impulse $P_E=mV_d$, which value is statistically average over all drifting electrons. In order to correctly calculate V_d value it is necessary to apply the momentum conservation law to this appeared transfer impulse $P_E=mV_d$. At that it should be taken into account that a growth rate of P_E is equal to electric field force eE, acting on each electron with its electric charge e, and a dissipation rate of P_E is equal to P_E/τ_p, where τ_p is relaxation time of P_E (according to accepted concepts of physics).

Then, at stationary electron drift, when growth and dissipation rates of P_E are equal, follows simple equation:

$$eE - \frac{P_E}{\tau_p} = 0. \tag{1}$$

From equation (1), which is dynamic balance for transfer impulse $P_E=mV_d$, follows well known expression for V_d:

$$V_d = \frac{eE}{m}\tau_p, \tag{2}$$

and also the expression for electron mobility (μ):

$$\mu = \frac{V_d}{E} = \frac{e}{m}\tau_p. \tag{3}$$

As clear from equations (2) and (3), both V_d and μ are proportional to τ_p value, which should be found in each particular case for correct calculation of V_d and μ values. Usually the value of τ_p is not known a priori, since τ_p value is defined by scattering (dissipation) mechanism of drifting electrons. Unfortunately, it has been long ago accepted (more than a century ago) to substitute desired τ_p value by the value of electron free path time τ, which may be easily calculated from known cross-sections of electron scattering in the substance. Thus, it is generally accepted to assume that $V_d=eE\tau/m$ (and $\mu=e\tau/m$). However, this substitution for the case of elastically (and isotropically) scattered electrons leads to significant underestimation of V_d (and μ) values, since in this case the τ_p value is much larger (from 16 to 4 times) than τ value. As was shown [7], this result follows from correct application of momentum conservation law to the electron transfer impulse $P_E=mV_d$, what is well confirmed by known experimental date for $V_d(E)$ dependences in dense rare gases [1-3].

It is reasonable to think, that τ_p should be equal to several values of τ, since dissipation (relaxation) of transfer impulse P_E may occurs only at electron scattering, whose statistically mean frequency is determined by τ. Thus, there should be quantitative relation between τ_p and τ values, i.e. it should be $\tau_p=Z\tau$, where Z is a number larger than 1. Then, according to equation (2), for value of V_d follows such expression:

$$V_d = \frac{eEZ\tau}{m}. \tag{4}$$

Equation (4) gives that the stationary drift impulse of electrons should be equal to:

$$P_E = mV_d = eEZ\tau. \tag{5}$$

In order to correctly calculate V_d value it is necessary establish Z value since the value of τ may be calculated. To obtain the value of Z, it is important to take into account real physics of stationary drift of elastically scattered electrons, i.e. to understand why constant value of V_d is achieved at applied constant E. For that it is convenient to consider transfer of electron assembly during one electron free path time τ, for which all drifting electrons should be scattered for certain. As clear, this scattering of electron assembly is isotropic since drifting electrons will be scattered on the atoms uniformly (in all directions). During time τ a statistically mean increase of $P_E=mV_d=eEZ\tau$ (in direction of E action) is equal to $eE\tau$. For stationary electron drift it must be so that a backscattered (against direction of E action) part of $P_E=mV_d=eEZ\tau$ during time τ must also be equal to $eE\tau$. According to equation (5) this backscattered part must be equal to Z^{-1} part of stationary electron drift impulse $P_E=eEZ\tau$. Thus, in order to obtain desired Z value it is necessary reveal what part of impulse $P_E=eEZ\tau$ is backscattered for the time τ. This part, which may be revealed by spatial analysis of P_E scattering [7], must be equal to Z^{-1}, what gives desired value of Z.

At spatial analysis of backscattered part of P_E it is necessary take into account the presence for drifting electrons of their thermal (isotropic) impulses, which introduce no contribution to P_E, but strongly affect its backscattered part, changing significantly Z value. Such spatial analysis of backscattered part of P_E was done for two limiting cases of E values [7]. For small E and small P_E, when drifting electrons are nearly thermal, it was obtained that Z=16. This value corresponds well to known Bohm coefficient 16, introduced by him to explain observed "anomalous" fast diffusion of thermal electrons in the gas of argon under crossed magnetic and electric fields. As was noted above, to explain "anomalous" fast electron diffusion, D. Bohm have used an idea (magnetohydrodynamic) of chaotic motion of electrons, what is true, but not necessary since the motion of drifting electrons in the gas substance is chaotic in principle due to isotropic scattering of drifting electrons (for each time τ in arbitrary directions). For large E values, when drifting electrons are strongly heated above their thermal energy and so their thermal impulses may be neglected, it was obtained that Z=4 [7]. This result for large E values, when only transfer impulse impulse P_E is scattered, may be substantiated very simply. As known, at isotropic scattering of some flow an exactly ¼ part of this flow is scattered in any direction, therefore in this case namely ¼ part of P_E will be backscattered, and consequently it means that in this case Z=4. As was shown [5,6], these theoretically obtained limiting values for Z and diminution of Z value from 16 to 4 at increasing E are entirely confirmed by known experimental data for electron drift velocity in dense rare gases.

4. Conclusions

The use of statistically correct theory, based on the law of momentum conservation for transfer impulse of drifting electrons $P_E=mV_d=eEZ\tau$, shows that the problem of "anomalous" fast transfer of electrons in tokamak plasma (and at known Bohm's diffusion) is caused by significant (from 16 to 4 times) underestimation of electron drift velocity (and mobility) at its accepted calculation, when it is believed that Z=1.

The law of momentum conservation for transfer impulse P_E gives that drift velocity (and mobility) of electrons in gas phase must be so fast really. Therefore, to suppress the damage of tokamak chamber by "anomalously" fast transfer of hot electrons, what disturbs conditions of controlled fusion realization, it is reasonable to increase inner diameter of tokamak chamber according to really fast electron mobility.

The damage of inner wall of tokamak chamber by fast transfer of hot electrons may be significantly suppressed due to additional heating of drifting electrons (above thermal energy of tokamak plasma) by additional electric field. As gives developed theory [1-3], such overheating lets diminish the "anomalous" fast transfer of electrons by several times (up to 4 times). As known, this possibility, revealed due to practical activities, is used already for suppression of "anomalous" transfer of hot electrons in tokamak (and stellarator) chambers.

It is important to take into account the established real properties of the forced transfer of electrons under the electric field (numerical relationship $\tau_p=Z\tau$) at investigation of substance (plasma) electron parameters based on measuring of electron conductivity. In particular, this allows correctly calculate the concentration of conduction electrons (based on the measured value of electron conductivity) on the base of true expression for V_d (calculated with relationship $\tau_p=Z\tau$), without leading to significant (in Z times) overestimation of electron concentration because of the generally accepted use of wrong relationship $\tau_\rho=\tau$.

References

[1] Boriev I A 2007 Some fundamental properties of electron transfer and heating in substance under electric field: the results of statistically correct theory and their confirmation by known experimental data *Proceed. Acad. Sci. ser. Energ.* **5** pp 106-14 (in Russian) https://elibrary.ru/item.asp?id=9584639

[2] Boriev I A 2006 Discovery of earlier unknown properties of electron transfer and heating in substance under electric field force and their vindification by statistical physics methods for open systems *XXXIII Int. Conf. Plasma Phys. and Controlled Fusion* February 13-17 2006 Zvenigorod Book of thesis p 188 (in Russian) http://www.fpl.gpi.ru/Zvenigorod/XXXIII/Lt/ru/BK-Boriev.doc

[3] Boriev I A 2005 Statistical basis of the theory of electron forced transfer (drift) and heating under electric field in the boltzman gas of atoms *Booklet of Scientific Council of RAS "Physics of low temperature plasma"* pp 116-121 (in Russian) http://ihed.ras.ru/council/booklets/2005/F_IEPHF.htm

[4] Gushchin E M, Kruglov A A, Obodovskii I M 1982 Electron dynamics in condensed argon and xenon *Journal of experimental and theoretical physics* **82** pp 1114-1125 (in Russian) Sov. Phys. JETP **55** pp 650-655 (in English)

[5] Boriev I A 2003 The explanation of the saturation effect of the drift velocity of excess electrons observed in condensed rare gases in the limit of a strong electric field *Russian Journal of Physical Chemistry B* **22** pp 80-87 (in Russian) https://elibrary.ru/item.asp?id=17300869

[6] Boriev I A 2003 On the reasons for the increase in the drift velocity of quasifree electrons in liquid inert gases when small amounts of molecular impurities are introduced *Russian Journal of Physical Chemistry B* **22** pp 103-110 (in Russian) https://elibrary.ru/item.asp?id=17300886

[7] Boriev I A 2012 Numerical relationship of the relaxation time of the electron transfer impulse in substance under electric field force with the electron free path time *Proceed. Acad. Sci. ser. Energ.* **3** pp 105-12 (in Russian) https://elibrary.ru/item.asp?id=17859204

[8] Boriev I A 2013 On the reasons of inconsistency of accepted approach for description of electron transfer and heating in substance under electric field force by solution of Bolzman's kinetic equation for electron velocity distribution function *XL Int. Conf. Plasma Phys. and Controlled Fusion* February 11-15 2013 Zvenigorod Book of thesis p 169 (in Russian) http://www.fpl.gpi.ru/Zvenigorod/XL/Lt/ru/FO-Boriev.doc

[9] Boriev I A 2014 Fundamental laws of classical and quantum physics follow from the features of microwave background radiation produced by dark matter seesaw motion *Int. J. Astron. Astrophys. and Space Science* **2** pp 7-11 http://www.openscienceonline.com/journal/archive2?journalId=703&paperId=1328

Solar wind as a consequence of Coulomb mirror effect in astrophysics

P I Vysikaylo and M A Korotkova

Moscow Region State University, Radio street 10A, Moscow, 105005, Russia

E-mail: filvys@yandex.ru

Abstract. The causes of the appearance of a solar wind consisting of protons, alpha particles and electrons arising as a result of ionization processes of atoms and molecules focusing to the Sun by gravitational forces are investigated. It is shown that a possible explanation of the occurrence of the solar wind is a slight violation of the neutrality of the Sun and the appearance of Coulomb mirrors for protons and alpha particles that reflect and accelerate light, positively charged particles. The parameters of the positively charged Sun are defined, allowing to estimate the reduced electric field in the photosphere, the chromosphere, the Sun's corona and to study the conditions necessary for reflection of positively charged particles from the positively charged Sun.

1. Introduction

Despite the huge amount of papers devoted to the problem of the origin and description of processes in the solar wind (SW) coming from the Sun to the Earth, at the present time there is no clear understanding of the causes responsible for the existence of the SW filling the heliosphere with the supersonic velocity which verified by numerous experimental observations.

Experimental research of SW began 58 years ago (on the Soviet rocket Luna-2 in 1959 and on American spacecraft Explorer 10 in 1961). The theoretical basis for understanding the dynamic processes of the formation of SW was laid by Parker in 1957 [1]. However, the mechanisms responsible for heating the base of the solar corona to temperatures of 1.5-2 million degrees and causing its expansion are still not entirely understood. Currently, the conventional models explain the heating of the solar corona and the existence of the SW by Alfvén waves [2,3,4]. It is believed that the mass composition of SW, as it moves, does not change in the interplanetary medium and therefore provides information on the chemical and ionic composition of the solar atmosphere. Heavy ions are formed in the lower corona, and when moving in the SW, their states do not change. Consequently, SW ions carry information about the conditions in the solar corona. Observations of SW are of practical importance, since the SW plasma is the main agent by which active processes on the Sun influence the state of near-Earth space and the Earth's magnetosphere, and, consequently, the weather and the well-being of the Earth's population.

Here, an entirely new mathematical model of SW is analytically studied, taking into account the weak violation of the neutrality (VN) of the mass of the entire Sun. In the opinion of the authors, the VN of the Sun is responsible (as in the ordinary single-pole discharge on a charged needle) for the escape and acceleration of positively charged ions and electrons from the Sun. The original results obtained on the basis of these calculations are given. It turns out that only knowledge of the types of ions established in experiments can change the calculated parameters of the Sun and SW under different conditions and allow us to investigate new physical mechanisms responsible for the formation of SW, the dynamics of its large-scale perturbations, as well as the mechanisms governing the comparative behavior of various ionic components for various types of SW flows.

In this problem, an important fact, verified by experimental observations with the help of satellites and spacecraft, is the reverse direction of the SW movement from the Sun [5], and not to the Sun, which would be understandable, as a result of gravitational attraction between neutral plasma structures. The solution of this paradox is possible only if it is assumed that the Sun and SW are positively charged [6, 7]. It has been established that the main carriers of mass in SW are protons and alpha particles [5].

As shown in [8], in plasma cumulative-dissipative (CD) structures, such as lightning, bead lightning, cathode spots, positive columns, striations, etc. processes of formation and prolonged retention of the positive charge by the entire CD structure may occur. The term CD structure is introduced for dissipative structures, in which the forces of cumulation (focusing) of energy-mass-momentum flows (EMMF) are significant [7]. The long-range Coulomb potential of any plasma structure is formed because of the greater mobility of electrons compared to ions or even protons. Therefore, a small part of the light-weight, in comparison with ions, high-energy electrons always leaves the plasma structure and thereby positive charges it. A positively charged structure tends to return the outgoing electrons. Some of the outgoing electrons with lower kinetic energy, indeed, return to the Coulomb potential well for the Fermi gas, but not all. Thus, turbulence (vortices) is formed in the region of charged plasma structures. Returning electrons, which receive additional kinetic energy from the electric potential, focus the entire positively charged structure in a single luminous whole by their pressure (Figures 1-4). These processes are responsible for the surface tension of plasma CD structures. Therefore, cylindrically focused lightning arises (Figure 4), stratified bead lightning (Figure 3), striations, 3D cathode spots, cylindrical and conical (Fig. 2) positive columns in experiments, etc. [9, 10]. The maximum electric field - E, - of a uniformly charged structure, as is clear from the solution of the Poisson equation, is achieved on its surface and is determined by the density of the charge in the structure that is not compensated by the electrons (they left the structure). Here, the reduced electric field or the parameter E/N is also maximal. Here N is the density of gas particles. Therefore, both the pressure and the average energy of the electrons compressing the positively charged structure are maximal there. Compression of the structure leads to heating of the entire electron gas in the structure, and the electron gas on the other hand counteracts its compression to the point. It is clear that to describe even weakly charged extended structures, such local parameters as the Debye radius are not very effective [11,12].

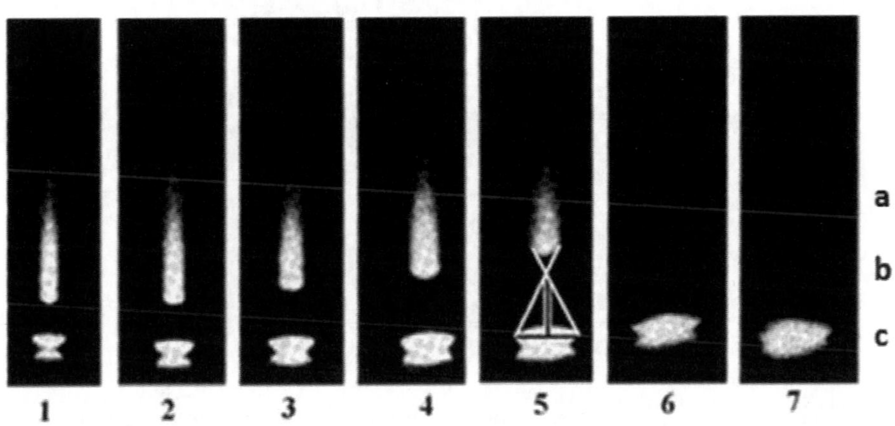

Figure 1. Dependence of the width of a cylindrical self-focusing discharge in nitrogen of special purity on current [13]: I = 0.6 (1); 0.8 (2); 1.1 (3); 1.65 (4); 2.2 (5, definition of the libration point); 2.9 (6); and 3.25 (7) mA; P = 5 Torr. a is a positive column (with a blue glow that diffuses towards the

anode, which is not photographed), b is Faraday space, c is a negative glow (or cathode spot, lower spot is a glare of the cathode spot on a mirror-polished electrode).

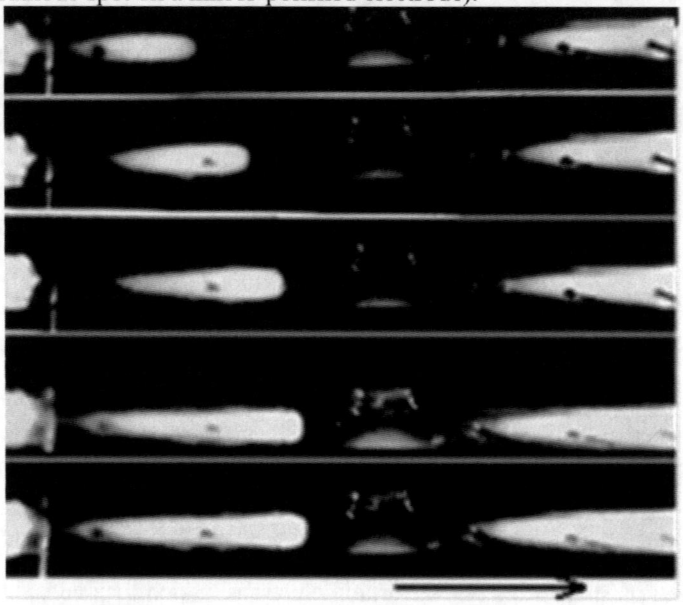

Figure 2. Structure of plasmoids in a discharge in a tube in nitrogen of special purity, depending on the discharge current at P = 15 Torr [14, 15]. The discharge is locally disturbed (the window size is 2 cm) in the center by a beam of high-energy electrons. (The gas was pumped at a velocity of 50 m/s from the cathode to the anode, along the arrow).

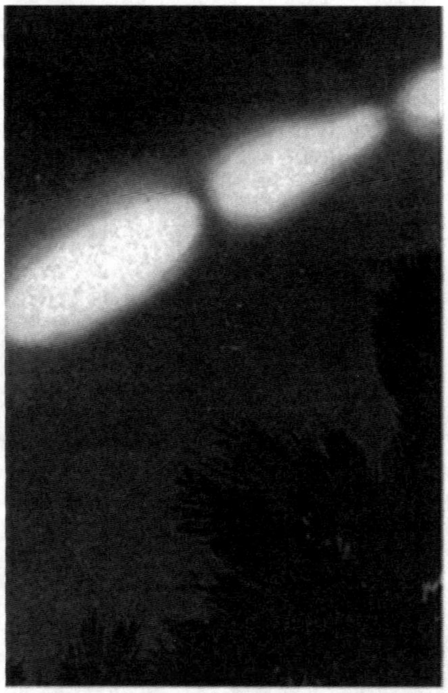

Figure 3. A bead lightning with internal points of libration between charged luminous structures on the background of a palm.

IOP Conf. Series: Journal of Physics: Conf. Series **996** (2018) 012015 doi:10.1088/1742-6596/996/1/012015

Figure 4. Discharge in air, with a persistent cross section. The glow indicates the radial focusing of the reduced electric field (E/N). This is an artificial lightning or plasma jet in which electron and ion fluxes are self-focused into a cumulative-dissipative structure.

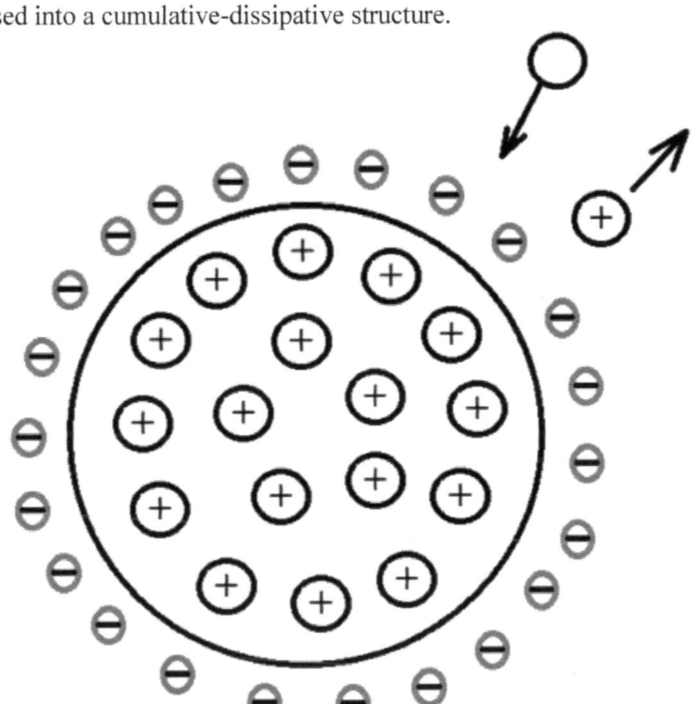

Figure 5. Model of the Sun as a positively charged structure.

With the help of this approach, the reverse motion of the cathode spot discovered by Stark in 1903, the 3D processes in Faraday space between the striations, the cathode spot and the positive (positively charged) column were explained and analytically calculated, libration points between positively charged structures were discovered [8, 9, 10, 16, 17]. These libration points (in Faraday spaces) between charged structures are analogs of Lagrangian points for gravitating structures, three of which were discovered by Euler 250 years ago in 1767.

On the basis of these representations, verified by a number of experiments, we will describe the "reverse" (against the gravitational forces) motion of SW, assuming that the Sun and SW, as plasma structures, have a positive charge. We shall characterize the energy of charged plasma particles in the chromosphere and solar corona by the parameter E/N. E is the electric field in the region of the charged Sun, and N is the particle density in this region.

2. Comparison of Coulomb and gravitational interactions between elementary particles

Since 1928, Einstein, Schrödinger, Eddington, Dirac, and others have attempted to create a unified theory of electromagnetic and gravitational fields [18]. For example, for two identical particles with masses m and charges e, the ratio of the electric potential Ke^2/r to the gravitational potential Gm^2/r is Ke^2/Gm^2, which for the electron gives a giant value of $4.17 \cdot 10^{42}$. In the 1930s, P. Dirac went further and noticed that the ratio of the electric force between the electron and the proton to the gravitational force between them is measured by a large number: $Ke^2/Gm_pm_e \approx 2.27 \cdot 10^{39}$. Here m_p, m_e are the masses of the proton and electron, respectively.

3. Modification of the Eddington' and Dirac' methods

In [6,10], the forces of gravity and Coulomb forces in objects with real dimensions from 10^{26} to 10^{-16} m were compared. In this problem, the ultimate goal was the same - the creation of a unified theory of interference of gravitational and Coulomb fields, not for elementary particles, but for real massive objects of the Cosmos and their parts (the centers of galaxies, clusters of galaxies, pulsars, quasars, planets, their atmospheres, etc.). In gravitating bodies, the mass is determined not by the mass of electrons, but by the mass of nucleons (protons and neutrons) that are the foundation of the ordinary nuclear matter from which the atom, molecule, any planet, star, etc. is built. The masses of nucleons are larger than the mass of electrons 1836 times. Therefore, we can neglect the mass of electrons leaving the structure in the total mass of the structure. According to the Coulomb's and Newton's laws in the gravitating objects of the Cosmos, for example, with an arbitrarily distributed charge along the radius, when comparing the forces of gravity and electric repulsive forces, it is necessary to compare the values of the forces $K \cdot (\alpha_{i1} \cdot N \cdot e) \cdot (n \cdot \alpha_{i2} \cdot e)/R^2$ and $G \cdot (N \cdot m_p) \cdot (n \cdot m_p)/R^2$. Here $K = 1/4\pi\varepsilon_0$, ε_0 is the electric constant, α_{i1} is the degree of uncompensated volumetric charge of protons in the entire CD structure, for example, a star; N is the total number of nucleons in the sphere of the structure of radius R, α_{i2} is the degree of uncompensated proton volume charge in the volume under investigation, for example, on the surface of the star at a distance R from its center; n is the number density of particles in this volume. The degree of uncompensated volumetric charge of the plasma structure - α_i is the ratio of the concentration of positively charged ions – n_i, which form the volumetric charge of the whole positively charged structure, to the density of the total number of nucleons – N, which determine the mass of the structure.

In this approach, the Eddington and Dirac numbers are modified into a number:

$$Ke^2/Gm_e^2 = 4.17 \cdot 10^{42} \rightarrow Ke^2/Gm_em_p = 2.27 \cdot 10^{39} \rightarrow K\alpha_{i1}\alpha_{i2}e^2/Gm_p^2 = 1. \quad (1)$$

The condition of gravitational and Coulomb forces equality leads to instability of the star, galaxy, their system or visible universe, analogous to the rotational 2D instability. The instability arising for the CD-structures of the Cosmos due to VN in the general case, for the convenience of classification, will be called the 3D instability of the Vysikaylo, in accordance with the terminology in [10]. The condition (1) allows us to determine:

$$((\alpha_{i1} \cdot \alpha_{i2})^{0,5})^* = \alpha_i^* = 0,9 \cdot 10^{-18}. \quad (2)$$

α_i* – the critical degree of the VN of the gravitating, charged body substance, when the body ceases to compress under the action of gravitational forces and, as in the case of rotational instability, parts of a charged **quasineutral** plasma in the form of a plasma, stellar, solar, etc. wind are emitted from its surface, as in the case of rotational instability [6, 10].

Vysikaylo' parameter of the dynamic order – $\alpha_i = (\alpha_{i1}\cdot\alpha_{i2})^{0,5}$ determines in the Cosmos the transition in CD structures from gravitational compression to Coulomb loosening or even sputtering and scattering of charged matter made of nucleons. The parameter α_i is determined by two parameters: α_{i1} is the parameter of the VN structure as a whole, and α_{i2} is the VN parameter of the element to be studied for the Coulomb instability, which is chosen in the charged CD structure. An element can have the value α_{i2} = 1, 1/2, 1/3 or less. In the case of the proton α_{i2} = 1, the Vysikaylo instability begins for such a mass with the VN parameter for the entire star for $\alpha_{i1} \geq 8.1\cdot10^{-37}$. At these values of α_{i1} and α_{i2}, the value of α_i will reach the Vysikaylo' limit, see equation (2).

4. Classification of charged stellar wind

Vysikaylo instability first of all begins for protons, since the parameter αi2 for them is the maximum and equal to 1. The values of $\alpha_i^2 = 2$ or 10 are impossible because of the impossibility of cumulation into the atomic nuclei of only protons. And for an atomic nucleus with 10 protons and 10 neutrons, the parameter $\alpha_i^2 = 1/2$. When the Vysikaylo parameter for the star $\alpha_{i1} \approx 8.1\cdot10^{-37}$ is reached, a reflecting Coulomb "mirror" automatically appears for the protons, and the free path to the surface of the charged star or to its interior for the proton will be temporarily closed because of the star's charge and the dominance of the Coulomb repulsion forces over gravitational attraction of protons. Positive charge of stars and Vysikaylo instability for protons, as the author supposes in [6,10], cause up to 90% of protons, experimentally observed in all energy-mass-momentum flows of the Cosmos. The possible range of the parameter α_i^2 values for cumulating on the attractor elements becomes clear and consists of $1 \geq \alpha_{i2} \geq 0$. The change in the parameters of ordinary space gravitating stationary objects extends from α_{i1} $\alpha_{i1} \approx 0.81\cdot10^{-18}$ to 0.

5. Calculations of positively charged Sun and solar wind parameters

As an example, which can verify the calculation of the charged stars parameters, we determine the condition under which protons are reflected from the positively charged Sun. Knowing the mass of the Sun or the number of nucleons in the Sun is $1{,}2\cdot10^{57}$ and the radius of the Sun is $R_S = 6{,}96\cdot10^8$ m, the VN parameter for the appearance of the Coulomb mirror reflecting proton in the Sun is $\alpha_{i1} \approx 0.81\cdot10^{-36}$. The solar charge in this case $q_S = e\cdot\alpha_{i1}\cdot N_S$ = 154 C and the electric field on its surface $E_S = Ke\cdot\alpha_{i1}\cdot N_S/R_S^2 = 2{,}86\cdot10^{-6}$ V/m.

The reduced electric field *E*/*N* (Townsend) for plasma particles (hydrogen atoms) significantly depends on the plasma particles density and rapidly grows from the photosphere to the chromosphere. We take the relation of *N* with the height from [19]. We calculate the relation *E*/*N* with the height.

It can be seen from Table 1 that in the photosphere region, where the concentration of plasma particles is large, the parameter E/N is small, but in the chromosphere this parameter grows rapidly with height and reaches significant values, which apparently leads to a substantial heating of the electrons already in this region of positively charged Sun. According to the above model, in the upper chromosphere and the corona, a non-equilibrium plasma is formed in the external electric field of the positively charged structure – the Sun with E/N values much larger than the breakdown fields in the completely ionized hydrogen plasma. This, as we suppose, generates SW in outer space and the removal of plasma, as in an ordinary plasmatron. Helium ions are observed in the SW. This means that all Sun's parameters are two or four times (see table 1 - E/N*) more. If we take into account the presence of the seven times ionized iron of Fe observed in the SW [20], this leads to an 9-fold increase in these parameters (E/N**), and the presence of xenon ions indicates the possibility of increase of these parameters up to 130 times, but this data already questionable.

Table 1. The values of N_H, cm^{-3} and E/N, Td as a function of height above the surface of the Sun

h, km / r, R_S	N_H, cm^{-3}	*E/N*, Td	*E/N* *, Td	*E/N* **, Td
0/1	$3.98 \cdot 10^{15}$	$7.19 \cdot 10^{-7}$	*E/N* x2 or x4	*E/N* x9
1 000	$3.16 \cdot 10^{13}$	$9.05 \cdot 10^{-5}$		
2 000	$6.31 \cdot 10^{12}$	$4.53 \cdot 10^{-4}$		
3 000	$1.99 \cdot 10^{12}$	$1.44 \cdot 10^{-3}$		
4 000	$7.94 \cdot 10^{11}$	$3.6 \cdot 10^{-3}$		
6 000	$2.51 \cdot 10^{9}$	1,14		
8 000	$1 \cdot 10^{9}$	2.86		
10 000	$6.31 \cdot 10^{8}$	4.53		
15 000	$1.99 \cdot 10^{8}$	13.78		
70 000	$7.94 \cdot 10^{7}$	29.75		
280 000	$1.26 \cdot 10^{7}$	115.79		
420 000	$5.01 \cdot 10^{6}$	223.31		
700 000/2	$1.58 \cdot 10^{6}$	452.28		
1 400 000/3	$3.98 \cdot 10^{5}$	798		
2 800 000/5	$6.31 \cdot 10^{4}$	1812		
6 200 000/10	$1 \cdot 10^{4}$	2942		
13 000 000/20	$2.51 \cdot 10^{3}$	2973		26757
44 000 000/65	$1 \cdot 10^{3}$	701		
150 000 000/215	$6.31 \cdot 10^{2}$	97.7		

6. Calculation of protons and alpha particles energy in the Earth's region

The Coulomb potential energy of a proton born at a distance *r* from the Sun will be:

$$W = K \cdot q_S \cdot e/r. \quad (3)$$

This potential energy can be completely changed to the kinetic energy of a charged particle far from the Sun if the particle does not experience inelastic collisions that substantially change its kinetic energy. For protons born in the Earth's orbit, $W = 1.5 \cdot 10^{-18}$ J (9.2 eV), where *r* is the astronomical unit (the distance from the positively charged Sun to the Earth) if the solar charge is 154 K.

From the experimental studies of the SW in the region of the Earth, it is known that the velocity of protons and ions in it reaches from V = 400 km/s to 1000 km/s [1]. Consequently, these protons came from a region close to the Sun. Knowing from the experiments the average speed of the solar wind near the Earth - V, we can estimate the kinetic energy of the protons in the solar wind:

$$W = m \cdot V^2/2 = 1.34 \cdot 10^{-16} \div 8.36 \cdot 10^{-16} \text{ J } (835 \div 5\,200 \text{ eV}). \quad (4)$$

Such energy can be obtained by protons born in the corona of the positively charged Sun and accelerated by its charge to such velocities in the region of the Earth. In this case, the mean free path, in which protons flying to the Earth, do not experience collisions with other nucleons, can be estimated from the formula:

$$L = 1/(\sigma N). \quad (5)$$

σ is the effective collision cross section of protons, which for energies of the order of 1 keV is equal to $2 \cdot 10^{-17}$ cm^2. In this case, protons born at distances of the order of 2 800 000 km to 6 000 000 km (5 ÷ 10 solar radii) will arrive to the Earth with a small number of collisions. At these distances, their potential energy at birth is about 3 keV, if the solar charge is of the order of 1000 K, i.e. protons, alpha particles and even six times ionized iron ions are reflected from the Sun (Figure 1).

If we consider that the Sun reflects alpha particles, iron ions, then we get an increase in the energy of positively charged particles in the region of the Earth also by two, four and nine times, respectively. Thus, experimental observations in the interplanetary space of the mass and charge compositions of the SW ions already provide valuable information not only about the solar atmosphere, but also about the average parameters of the Sun, such as its total charge, the electric field magnitude in the corona, and make a significant contribution to the verification of mathematical models of processes on the Sun.

Conclusion

The paper first proposed a specific mechanism for heating the solar wind plasma in its corona in the constant Coulomb field of the entire Sun, as a positively charged cumulative-dissipative structure. Cumulative processes on the Sun are caused by gravitational forces, and dissipative processes are determined mainly by the scattering of electromagnetic radiation and to a lesser extent by "corpuscular rays" - the solar wind. As shown in [21], the cumulative processes occurring in the gravitational or Coulomb potentials with the U (r) ~ 1/r law can only convert half of the potential energy to the kinetic energy of the particles remaining in the CD structure. The second half of the potential energy must be scattered into the space surrounding the cumulative structure. This follows from the virial theorem [22] or Newton's third law [21]. In excited Coulomb structures such as atoms, molecules, the take-off of ½ potential energy is carried out by light quanta – electromagnetic waves. In more complex structures, for the fulfillment of this law of cumulation, streams of particles with non-zero mass, for example, electrons, can be used, and after charging the CD structure with positive charge, protons and other ions. According to Parker [1], the first clear statement that something but light comes to the Earth from the Sun was expressed in 1896 by Norwegian physicist Kristian Birkeland. He came to this conclusion on the basis of the fact that the auroras are very similar to the electric discharge in the newly invented tubes, which generate charged particle fluxes ("cathode rays"). Thus, the energy for the outflow of the solar wind is due to gravitational compression of the Sun in its own gravitational field. This is accompanied by thermonuclear reactions of the neutronization of hydrogen atoms to helium atoms, the release of a small fraction of electrons from the Sun, as the most mobile gas, the charging of the entire Sun to several hundred Coulombs, the formation of Coulomb mirrors reflecting positive ions and the formation of solar wind from the Sun. We can say that the authors of this work are followers of the ideas of K. Birkeland and R. Clausius, who introduced the concept of the virial in 1870.

Our analysis using mathematical methods of the causes of the solar wind shows that an indirect sign of cumulative processes in CD structures such as the Sun and the Solar System can be not only rotation, but also a violation of neutrality and the formation of oppositely directed bicyclic rotating flows that cumulate the entire mass of the Sun to its center and defocusing mass of the solar wind leaving the solar system, see [7].

As proved by analytical calculations, having determined in the experiments the ion composition of the solar wind and finding the minimum value of the ratio Z/m, according to the method proposed in this paper, it is possible to calculate the minimum electrical parameters of the Sun itself, its corona and analytically calculate the energy parameters of the solar wind in the region of the Earth. This requires detailed experimental studies of the parameters of the solar corona (the particle density profile), and reliable data on the ion composition of the solar wind arising from Coulomb mirrors reflecting these positively charged ions from the Sun.

The smaller the Z/m ratio, the greater the positive charge of the Sun for the equality of Coulomb repulsive forces and gravitational attraction forces for a given ion. The ion velocity in the solar wind is also essentially dependent on this parameter and on the place where the ion was born. This determines the variance of the solar wind velocity from the ratio Z/m.

References

[1] Parker, Eugene Newman. Solar Winds. No. SCR-126. Chicago. Univ., 1959.

[2] McIntosh, S.W., de Pontieu, B., Carlsson, M., Hansteen, V., Boerner, P., Goossens, M.: 2011, Alfvénic waves with sufficient energy to power the quiet solar corona and fast solar wind. Nature 475, 477.

[3] Artemyev A.V., Zimovets I.V., and Rankin R., Electron trapping and acceleration by kinetic Alfvén waves in solar flares. (2016) A&A 589, A101.

[4] Artemyev A.V., Rankin R., and Vasko I.Y., Upper limit of electron fluxes generated by kinetic Alfvén waves in Maxwellian plasma. (2016) J. Geophys. Res., 121, 8361–8373

[5] Podgornyi I.M., Sagdeev R.Z. "Physics of interplanetary plasma and laboratory experiments" Sov. Phys. Usp. 12 445–462 (1970)

[6] Vysikaylo P.I. Neustoychivost' fokusiruyushcheysya massy. Mezhdunarodnaya konf. MSS-09 «Transformatsiya voln, kogerentnye struktury i turbulentnost'» 23-25 noyabrya 2009 g. Sbornik trudov. – M.: LENARD, 2009. – 512 (s. 387).

[7] Vysikaylo P.I. Arkhitektura kumulyatsii v dissipativnykh strukturakh. Saarbrucken: Palmarium Academic Publishing. 2013, - 352 c.

[8] Vysikaylo P.I. Periferiynye kulonovskie sily, klassicheskie i kvantovye membrany fokusiruyushchie plazmoidy (obzor) //Uspekhi prikladnoy fiziki. 2015. tom 3, № 5. S.471-478.

[9] Vysikaylo, P.I. Surf. Engin. Appl.Electrochem. (2013) 49: 222. https://doi.org/10.3103/S1068375513030125

[10] Vysikaylo P.I. Surf. Engin. Appl.Electrochem. (2012) 48: 212. https://doi.org/10.3103/S106837551203012X

[11] Vysikaylo P.I. Skachki parametrov neodnorodnoy stolknovitel'noy plazmy s tokom obuslovlennye narusheniem kvazineytral'nosti. // Plasma Physics Reports, 1985, t. 11, № 10, s. 1256-1261

[12] Vysikaylo P.I., Tsendin L.D. Rezko neodnorodnye profili kontsentratsii plazmy v razryade pri povyshennykh davleniyakh. Plasma Physics Reports. 1986, 12(10), 1206–1210

[13] Vysikaylo P.I., Glova A.F., Smakotin M.M. Statsionarnyy tleyushchiy razryad v azote s otritsatel'noy vol't-ampernoy kharakteristikoy. Plasma Physics Reports. 1988, 14(6), 734–736.

[14] Babichev V.N., Vysikaylo P.I, Pis'mennyy V.D. i dr. Eksperimental'noe issledovanie ambipolyarnogo dreyfa plazmy, vozmushchennoy puchkom bystrykh elektronov. Dokl. AN SSSR. Fizika. 1987, 297(4), 833–836.

[15] Babichev V.N., Vysikaylo P.I., Golubev S.A. Eksperimental'noe podtverzhdenie sushchestvovaniya skachkov parametrov gazorazryadnoy plazmy. Technical Physics Letters. 1986, 12(16), 992–995.

[16] Vysikaylo P.I. Kumulyativnaya fizika kristallov i plazmoidov. // Uspekhi prikladnoy fiziki. 2015. tom 3, № 3. S.226-235

[17] Vysikaylo, P.I. J. Exp. Theor. Phys. (2004) 98: 936. https://doi.org/10.1134/1.1767560

[18] Born, Max, Roger John Blin-Stoyle, and John Michael Radcliffe. Atomic physics. Courier Corporation, 1989.

[19] Martynov D.Ya. Kurs obshchey astrofiziki. M.: Nauka, 1971. 616 s.

[20] Ermolaev Yuriy Ivanovich «Eksperimental'noe izuchenie krupnomasshtabnoy struktury solnechnogo vetra» (Spetsial'nost' Fizika Solntsa). Dissertatsiya na soiskanie uchenoy stepeni doktora fiziko-matematicheskikh nauk. Moskva 2002 g.

[21] Vysikaylo P.I. «Kvazikuperovskie» bitsiklony. Turbulentnye struktury s vrashcheniem i kumulyativnymi struyami. // Inzhenernaya fizika 2013. № 7, c. 3-36.

[22] Landau, Lev Davidovich, and Evgenii Mikhailovich Lifshitz. Course of theoretical physics. Elsevier, 2013.

IOP Conf. Series: Journal of Physics: Conf. Series **996** (2018) 012016 doi:10.1088/1742-6596/996/1/012016

Model of superconductivity formation on ideal crystal lattice defect–twin or twin boundary (MSC-TB)

V A Chizhov[1], F S Zaitsev[2], V L Bychkov[3]

[1]LLC «Processing of energy materials»,
[2]M.V. Lomonosov Moscow State University,
[3]NIISI RAS

E-mail: bychvl@gmail.com

Abstract. The report provides a review of the experimental material on superconductivity (SP) accumulated by 2017, a critical analysis of the Bardeen-Cooper-Schrieffer theory (BCS) has been given, and a new model of the super-conductivity effect proposed in works of V.A. Chizhov has been presented. The new model allows to understand the mechanism of the SP formation and to explain many experimental facts on the basis of the theory of pro-cesses occurring in the ideal defect of the crystal lattice – the twinning boundary (MSC-TB). Specific materials, including new ones, are described, which, in accordance with the theory of MSC-TB, should have improved properties of SC, promising directions for further research are formulated.

1. Introduction

More than 100 years have passed since a discovery of superconductivity (SC) in mercury at a temperature of T_c = 4.2 K, and a secret of a surprising effect of the SC remains unclear. A confirmation of this is a discovery of high-temperature superconductivity (HTSC) in Cuprate ceramics in 1986: Bednorz and K. Mühler, conducting research in the Zurich branch of IBM, found that ceramics based on copper, lanthanum and barium oxide at T_c = 30 K transforms into a superconducting state. This message shocked the whole scientific world. In 1987, a similar compound of yttrium barium Cuprate showed a transition to the superconducting state above the temperature of liquid nitrogen T_c = 92 K. New perspectives of superconductivity were opened, since the cost of liquid nitrogen is much cheaper than of liquid helium.

In 2005, the SC was discovered in ferrate-containing compounds – (LaOFeAs) Tc = 4K, where nobody expected the appearance of SC, because it was believed that ferromagnetism and superconductivity were incompatible. In 2008, the partial replacement of oxygen O2 by fluorine F (La{O1-xFx}FeAs (x=0.05-0.12)) led to an increase in the critical temperature by several times Tc = 26 K. The reason for raising of Tc remains unclear.

The following superconductivity theories are well known: London's (1935); Ginzburg-Landau (1950); Bardeen-Cooper-Schrieffer (BCS, 1957). See the bibliography in [1, 2]. By now, the BCS theory is widely accepted. However, none of these theories predicted the experiments on the SC, and cannot fully explain them and predict the origin of the SC in new materials. In fact, all the discoveries of the SC, starting from the very first in 1911, are a matter of chance. The same trend continues to the

IOP Conf. Series: Journal of Physics: Conf. Series **996** (2018) 012016 doi:10.1088/1742-6596/996/1/012016

present. The search for new superconducting materials is intensively continuing, but there is no systematic approach.

The existing theories of the SC also cannot explain a whole line of experimental facts. The absence of a SC in very good electro-conductive materials (gold, silver, platinum, and copper), and at the same time copper presence in many HTSC compounds – Cuprates. There is no complete description of the Meissner-Oxenfeld effect mechanism. Only the expulsion of the magnetic field from the volume of the superconductor or levitation – "the coffin of Mohammed" is indicated. However, in reality, in this effect, a levitation of a permanent magnet over a superconductor, and attraction, and a mirror, and memory from a previous exposure to an external magnetic field are observed simultaneously. There is also no understanding of the mechanism of superconductor degradation.

The main reason for the absence of convincing explanations of experiments using the BCS theory is, apparently, non-fulfillment of conditions for the applicability of this theory in reality. The BCS theory was developed for ideal crystals, but it is impossible to use such crystals in experiments. In addition, in the BCS theory, the distances at which Cooper pairs have to interact appear to be much greater than the typical size of the crystal lattice. There are many structural elements of the crystal between the pairs (point donors, acceptors, vacancies, dislocations, small-angle boundaries, twins), but in the BCS theory an effect of such elements on the interaction of pairs is not taken into account.

2. Short description of the model of SC origination on twin boundary MSC-TB

In practice, it is impossible to achieve creation of an ideal crystal without defects. All superconductivity was found in polycrystals. In experiments on SC, one usually works with polycrystals. Polycrystals always have ideal crystal lattice defects-twins or twin boundaries (TB), see Fig. 1-3. Even specially grown high-purity single crystals with SC properties have residual impurity content greater than 1014 cm-3, which can create energy wells of twin boundaries (TB) type.

Let us consider the flow of quasifree electrons of the metallic phase in the TB. We introduce the notation: ΔETB is the energy of the energy well of the twin boundary with respect to the energy of the crystal Ecr; Wke is the kinetic energy of the electron; Eph is the energy of phonons; T is the temperature.

We represent the "work" of the twin boundary in the dynamic mode as $\Delta ETB = \partial ETB / \partial T$. The energy of the "well" of the twin boundary $|\Delta ETB|$ increases with decrease of temperature T and becomes greater than the kinetic energy of the electron Wke. The electron is captured by the "well" and is thrown to the surface, because a direction of movement outwards is more energetically favorable. This action occurs not with one electron, but with all that satisfy the capture condition $|\Delta ETB| \geq Wke$, where Wke = meVd2 / 2, see Fig. 2, 3. One should note an influence of the surface energy Esur on this process, which should allow electrons to "exit" from the TB, i.e. the condition $Esur < \Delta ETB < Ecr$ has to be satisfied for Esur (for more details, see [1, pp. 37-43]).

Figure 1. A photo of the crystal with the twin boundaries.

Electrons ejected from the TB almost without hindrance go back to the volume of the crystal because of their excessive density above its surface, but under the action of the same TB they are again ejected to the surface.

Thus, in the near-surface layer by means of the TB and its energy well |ΔETB| a vortex current is originated, and, consequently, a magnetic field rot $\mathbf{B} = \mathbf{j} + \partial\mathbf{D}/\partial t$, see Fig. 3, 4. The effect of the TB on the electron system in a crystal with lowering of its temperature is described in more detail in [1, p. 55-121].

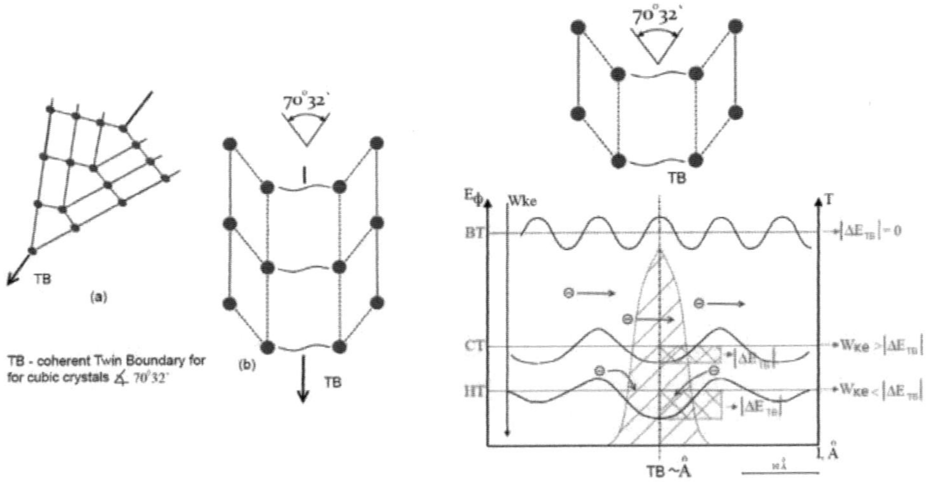

Figure 2. Scheme of a Twin-Boundary and its energy influence on the electron system in a crystal

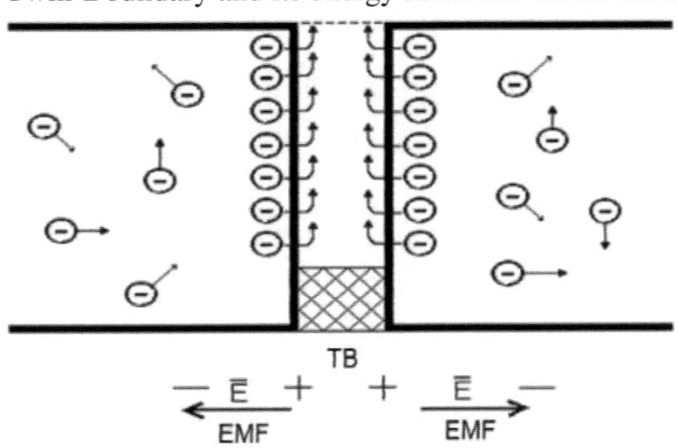

Figure 3. Scheme of electrons capture by the Twin Boundary.

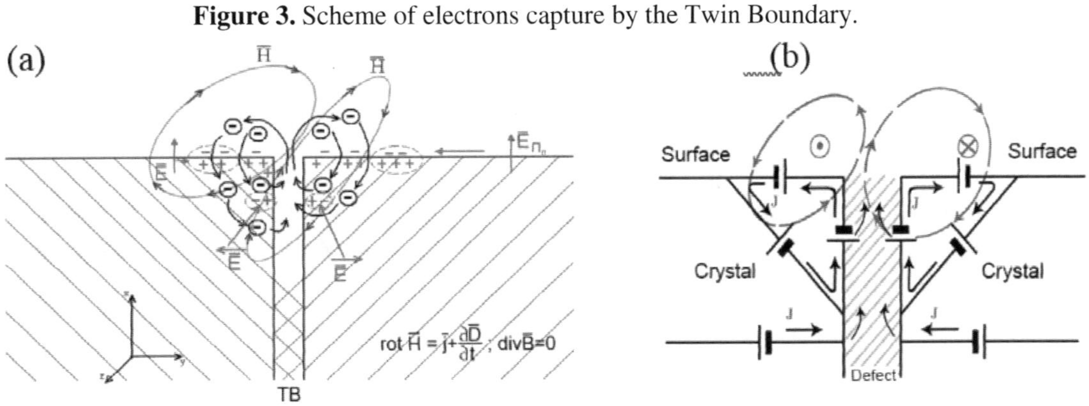

Figure 4. (a) – scheme of vortex currents and magnetic field origination by the TB. (б) – equivalent electric scheme.

3. SC effect and the Meissner-Oxenfeld effect

MSC-TB allows to explain many experimental facts [1, 2]. Let us stop here only on a brief explanation of the effect of the SC ("absence" of resistance) and the Meissner-Oxenfeld effect. Let us take into account that there are many twin faces in a polycrystal. Estimates [1, 2] show that for 1 cm-2 there are more than 108 twin boundaries, that is, NTB ≥108-1010 cm-2 emission channels for electrons, see Fig. 5.

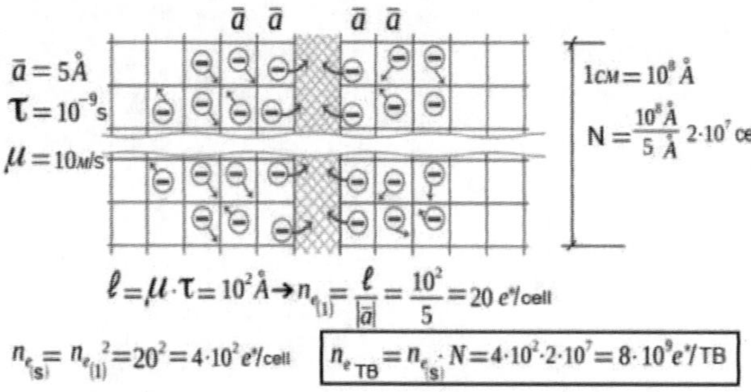

Figure 5. Parameters of electron capture of one TB and estimate of TB density per 1cm-2, ā is geometric parameter of the crystal lattice, N is the number of cells in 1 cm of the twin boundary face, e* is quasi free electron, μ is the electron drift velocity, τ is relaxation time, neTB is the number of electrons captured by TBF during the relaxation time τ.

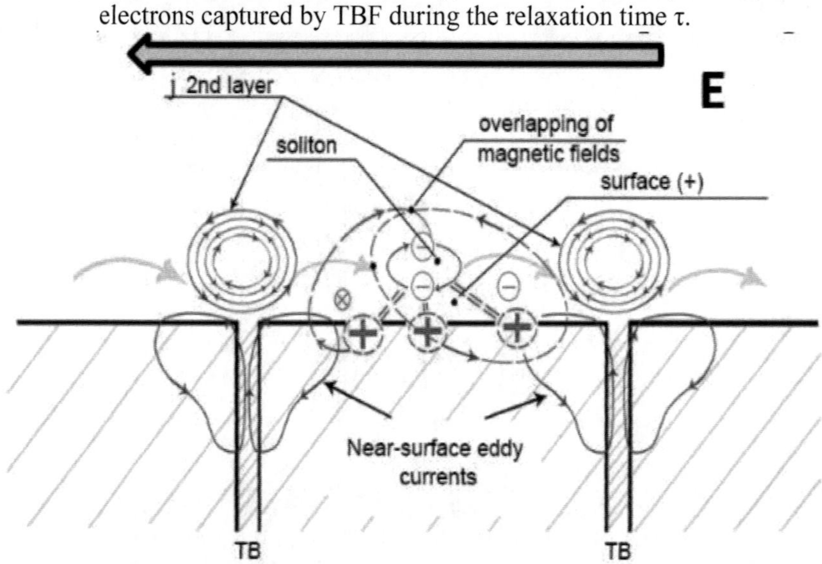

Figure 6. Scheme of electronic exchange between near-surface vortex currents created by the TB. The excited vortex of one TB transmits a charge to the unexcited vortex a charge with a jump, with formation in it of a quantum of a magnetic flux – a fluxoid.

If the condition for overlapping of near-surface vortex currents is satisfied at the lowering of the temperature, see, for example, Fig. 15 in [2], and an elec-tric or vortex magnetic field is applied to the crystal, then practically no-obstacle exchange of electrons occurs between the vortices over the TB. The effect of SC takes place, Fig. 6.

The Meissner-Oxenfeld effect in the MSC-TB model is explained by the excitation and orientation of additional vortex currents near the TB when the external magnetic field is applied, which, because of the presence of the TB, flow then for "infinitely" long time. This convincingly explains the properties of levitation, attraction, mirror and memory. More detailed analysis of these effects is presented in [2, p. 18-27].

IOP Conf. Series: Journal of Physics: Conf. Series **996** (2018) 012016 doi:10.1088/1742-6596/996/1/012016

Conclusion

The proposed model of MSC-TB is based on the analysis of data from all major experiments on SC and uses the most basic physical concepts and laws. There is neither a first-order SC nor the second-order SC, nor a SC of the 1.5-th kind, neither bosons made of electrons nor the polarized phonons. According to the MSC-TB model, in order to obtain the superconductivity effect (SC), it is necessary to create: 1) quasifree electrons of the metal phase type; 2) conduction channels on the crystal surface of the twin-boundary type; 3) the conditions for the capture of electrons into the twin boundaries and their ejection onto the surface; 4) stability of emission channels to degradation. Adequate understanding of the SC mechanism allows predicting superconducting materials with improved properties.

We can formulate what the MSC-TB model predicts for the creation of new SC materials. It is necessary to search for crystals with strong interatomic bonds – they are alkaline-fluoride systems of the type (d (x) + f (y) + MgF (z-δ)), (... CaF), (...... .CsF), (...... .RbF) with large lattice period ā; – in crystals with strong ionic-covalent bond – they are chalcogenides of metals; – in crystals with strong covalent type – such as Diamond.

References

[1] V A Chizhov, Again about superconductivity or experiments are awaiting answers. Moscow: "Sputnik" Publisher. 2015, (in Russian)

[2] V A Chizhov, Again about superconductivity or experiments are awaiting answers. Part II. Moscow: "Sputnik" Publisher 2017, (in Russian)

IOP Conf. Series: Journal of Physics: Conf. Series **996** (2018) 012017 doi:10.1088/1742-6596/996/1/012017

Existence of dark matter with observed properties of cosmic microwave background radiation substantiates three conservation laws of classical physics and all principles of quantum mechanics as creates the value of Planck's constant

I A Boriev

The Branch of Talrose Institute for Energy Problems of Chemical Physics of Russian Academy of Sciences, Chernogolovka, 142432, Russia

E-mail: boriev@binep.ac.ru

Abstract. Astronomical data indicate a presence of dark matter (DM) in the space, what is necessary for explanation of observed dynamics of the galaxies within Newtonian mechanics. DM, at its very low density ($\sim10^{-26}$kg/m^3), constitutes main part of the matter in the Universe, 10 times the mass of all visible cosmic bodies. No doubt, namely properties of DM, which fills space, must determine its physical properties and fundamental physical laws. Taking into account observed properties of cosmic microwave background radiation (CMBR), whose energy is ~90% of all cosmic radiation, and understanding that this radiation is produced by DM motion, conservation laws of classical physics and principles of quantum mechanics receive their materialistic substantiation. Thus, CMBR high homogeneity and isotropy ($\sim10^{-4}$), and hence the same properties of DM (and space) justify momentum and angular momentum conservation laws, respectively, according to E. Noether's theorems. CMBR has black body spectrum at ~2.7K with maximum wavelength $\sim1.9\cdot10^{-3}$m, what allows calculate the value of mechanical action produced by DM thermal motion ($\sim7\cdot10^{-34}$ J·s). This value corresponds well to the Planck's constant, which is the mechanical action too, what gives materialistic basis for all principles of quantum mechanics. Obtained results directly confirm the reality of DM existence, and show that CMBR is an observed display of DM thermal motion. Understanding that namely from DM occur known creation of electron-positron pairs as contrarily rotating material vortexes (according to their spins) let substantiate positron nature of ball lightning what first explains all its observed specific properties.

1. Introduction

Three fundamental laws of classical physics (conservation laws of momentum, angular momentum and energy) and also known principles of quantum mechanics (Planck`s constant, Heisenberg`s uncertainty principle, de Broglie`s wave mechanics, etc) have no physical (materialistic) vindication of their origin. All they are not derived from any higher order laws (principles), but nevertheless they are used with confidence since they are in good agreement with experimental data. However, they do not contradict to known experimental data only if they are used in the different domain of physics (classical or quantum). Thus, there are grave problems in fundamental physics: revealing of the reason of strongly different description (classical or quantum) of physical processes in the nature and of physically inexplicable idea of corpuscular-wave dualism. These problems demonstrate the incompleteness of understanding of the real reasons of classical physics laws and quantum mechanics principles and point out the necessity for hunting the reasons of their genuine origin.

No doubt, that all fundamental laws of classical physics and the principles of quantum mechanics must have their materialistic substantiation on the base of observed properties of the nature. Actually, as may be shown [1, 2], now accumulated data about observed properties of the cosmic space let give such substantiation by using the recently (1965) established properties of cosmic microwave background radiation (CMBR) [3] and taking into account the existence of dark matter (DM), which fills all ambient space [4-7]. Given substantiation is based on reasonable assumption that observed well known properties of CMBR, which also exists in all ambient space, is produced by certain motion of DM, which should has the same space properties as CMBR.

Really, observed high spatial homogeneity and isotropy of CMBR and, naturally, the same properties of its producing DM, validate conservation laws of momentum and angular momentum, correspondingly, in accord with known E. Noether`s theorems (1918) [8, 9]. These mathematical theorems assert that to each conservation law of physics match up the certain property of space, particularly, conservation laws of momentum and angular momentum will hold true only if the space is, correspondingly, highly homogeneous and highly isotropic.

Besides, as shown [1, 2], CMBR emitting properties (blackbody spectrum at temperature T~2.7K with maximum wavelength $\lambda \approx 1.9 \cdot 10^{-3}$m) let obtain the kinetic energy of producing this spectrum an equilibrium thermal motion of DM (with the use of Boltzmann`s constant). From the value of kinetic energy of such DM motion (obviously with the speed of electromagnetic waves $c=3\cdot 10^{8}$m/s) an action function value for DM seesaw motion may be calculated. Obtained value of this action function ($\sim 7\cdot 10^{-34}$J·s) is well equal to Planck`s constant $h=6.6\cdot 10^{-34}$ J·s, which really is an action function too. This result discovers materialistic origin of Planck`s constant, as the result of DM thermal motion, and gives, obviously, materialistic base to Heisenberg`s uncertainty principle and to others conceptions of quantum mechanics, which all follows directly from the existence of Planck`s constant [10].

As clear, such motion of DM must strong disturb a spatial motion of light elemental particles (like electron) what may be correctly described by the methods of quantum wave mechanics [10], proposed by L. de Broglie. But DM motion could not disturb significantly a spatial motion of much heavier particles (atoms, molecules and so on), so their spatial motion may be well described by the laws of classical physics. Thus, DM thermal motion explains the logical necessity of different classical or quantum description of physical processes in the nature and the reason of used corpuscular-wave dualism approach. Simply speaking, spatial motion of light elemental particles (like electron, etc) undergo to strong influence from DM, and their disturbed motion may be well described by wave mechanics with the use of Planck`s constant produced by DM motion. But, when such light particles interact (encounter) with another particles, they act as a corpuscular, which they really are.

Taking into account the existence of DM and reasonable comprehension, that electron-positron pairs (both of which is stable elemental material particles) are created from DM as material vortexes (according to their spins), let substantiate positron nature of mysterious atmospheric phenomenon – ball lightning (BL). As shown [11, 12], understanding of BL nature as a bunch of big amount of positrons allow first explain all observed specific properties of BL, particularly, its ball-shaped luminescence, visible size and life time, which are determined by behavior in the air of X-ray radiation from annihilation of BL positrons with electrons of the air molecules.

2. Substantiation of Fundamental Conservation Laws of Classical Physics

Well known fundamental laws of classical physics (momentum, angular momentum and energy conservation laws), which are used for description of observed physical processes in nature, have not its certain substantiation. Another words, the reason of their performance is not established exactly, and besides, what is important, the bound of applicability of these laws is not established too.

Known (from 1918) E. Noether's mathematical theorems denote conditions for realization of these three fundamental laws, according to them ambient space must exhibit right properties. Momentum conservation law takes place if space is uniform, angular momentum conservation law takes place if

IOP Conf. Series: Journal of Physics: Conf. Series **996** (2018) 012017 doi:10.1088/1742-6596/996/1/012017

space is isotropic, and energy conservation law takes place if trend of physical processes is independent on its starting moment. As clear, in order to these fundamental laws could really take place in the nature, the space must obeys the properties stated by E. Noether`s theorems.

Usually the applicability of conservation laws of classical physics is based simply on assumption that ambient space is empty and possess the properties of Euclidian geometry (the spatial homogeneity and isotropy), what accords to conditions of E. Noether`s theorems. Indeed, resent cosmological data indicate that real space is not empty, since it is filled by DM [4, 7], which existence in cosmic space was suggested long ago by classical astronomy to explain observed dynamics of galaxies according to Newtonian mechanics. As may be shown [1, 2], namely the properties of DM with the observed properties of CMBR, which is producing by DM motion, determine the applicability of conservation laws of classical physics and known principles of quantum mechanics.

As discovered in 1965 by Arno Pensias and Robert Wilson, the observed from the Earth a weak enough (like a noise) cosmic microwave radiation are characterized by high isotropy and a temperature about ~3K (for that they received Nobel Prize in physics in 1978). Now is well established, that this radiation, which is known as CMBR, is the most powerful among the all detected radiations from the space (about 90% of their total energy) and is characterized by very high isotropy and also by high homogeneity. Besides, as is established, CMBR has the blackbody spectrum at the temperature ~2.7K and the maximum of its radiation wavelength at $\sim 1.9 \cdot 10^{-3}$m.

It is clear that observed high isotropy and high homogeneity of CMBR mean that its producing DM has the same properties (in large space scale). Hence the ambient space, filled by DM, is also highly isotropic and highly uniform, what according to E. Noether`s theorems give real materialistic base for realization in the nature of the fundamental laws of the angular momentum and the momentum conservation, correspondingly. As to fundamental law of energy conservation, it takes place, according to E. Noether`s theorem, if run of observed physical processes is not depend on their starting moment, what means "time" constancy (or uniform flow of "time"). Such condition is well consistent with observed independence of behavior (on starting moment) of all physical processes in our materialistic nature. In this connection it should be understand that the mankind always use for metering the "time" flow any other physical processes, which are used as a "reference clock".

It is well known that as a "reference clock" mankind has used in its long history many very different materialistic processes taking place in the nature. Now as a "reference clock" an "atomic clock" is used, which is based on the frequency of atomic radiation (that also means the use of real materialistic process). Therefore, it is clear, that kinetic properties of all physical processes taking place in the nature come to know by mankind due to their comparison with any other physical (materialistic) process. No doubt, such understanding means that in nature really there is no "time": neither absolute "time" no relative "time". The overall using by mankind the concept of "time" is simply a very convenient parametrical approach to describe observed kinetics of physical processes in our materialistic nature (no more, then that).

As a resulting important conclusion, it must be underlined that fundamental laws of classical physics operate well in the nature not because of the E. Noether`s theorems exist by itself, but since the real properties of ambient space, which is filled by DM, are in accord with conditions of E. Noether`s theorems.

3. Materialistic Base of Known Principles of Quantum Mechanics

As well established CMBR has the blackbody spectrum at temperature ~2.7K with maximum wavelength $\sim 1.9 \cdot 10^{-3}$m. Taking into account the existence of DM which fills all the space, it is reasonably assume that observed CMBR is produced by equilibrium thermal motion of this DM. Such materialistic point of view, which let substantiate the applicability of fundamental laws of classical physics, as it is shown above, no doubt must also reveal the materialistic reason of quantum mechanics and its principles, including the known value of Planck`s constant.

The fact that CMBR has blackbody spectrum at $T \approx 2.7$K demonstrates that DM, which creates the CMBR, is in equilibrium seesaw motion at this temperature. Taking into account that DM is really in equilibrium motion, the value of kinetic energy of DM seesaw motion (ε) should be equal to DM thermal energy, which may be obtained from DM temperature with the use of the universal Boltzmann`s constant $k=1.38\cdot10^{-23}$J/K, which connect temperature of equilibrium thermal motion of some parts of any matter with mean-square velocity of motion of these parts. In the case of DM seesaw motion, the velocity of this motion is obviously equal to the electromagnetic wave speed ($c=3\cdot10^{8}$m/s) since these waves propagate namely in DM filling all ambient space. Obtained value of DM kinetic energy is $\varepsilon=3kT/2=5.6\cdot10^{-23}$J that let estimate the mean mass of these matter parts, namely, an effective mass of DM (m), which produces such DM seesaw motion: $m=2\varepsilon/c^2\sim1.2\cdot10^{-39}$kg.

It is clear that observed maximum wavelength of CMBR ($\lambda\approx1.9\cdot10^{-3}$m) is well equal to the mean value of the amplitude of DM seesaw motion, and that this amplitude corresponds to the damping length of DM seesaw motion. It is also clear that λ means the damping length for DM effective mass momentum (p), which is equal to $p=m\cdot c$. The established parameters of DM seesaw motion (the effective mass m, its motion speed c and the damping length λ for p) let estimate the value of mechanical action (an action function) for such DM seesaw motion by two known manner (from conceptions of classical mechanics and of quantum mechanics).

According to conception of classical mechanics an action function value (S) is a measure of physical motion and is equal to the product of energy and a "time" of this energy existence (dissipation). Taking into account that full circle path of DM seesaw motion energy is equal to 2λ then its "time" is equal to $2\lambda/c\approx1.3\cdot10^{-11}$s. So it may be obtained that S value for DM seesaw motion is equal to

$$S=2\varepsilon\lambda/c\approx5.6\cdot10^{-23}\cdot1.3\cdot10^{-11}\text{J}\cdot\text{s}\approx7.3\cdot10^{-34}\text{J}\cdot\text{s}.$$

As it is seen, such classically obtained estimation of the value of an action function of DM seesaw motion corresponds well to the Planck`s constant h, which physically is an action function too.

As it usually accepted in quantum mechanics the S value should be obtained by multiplying the mechanical momentum and its damping length. In the case of DM seesaw motion its mechanical momentum is equal to $p=m\cdot c$ and damping length of this momentum is equal to λ. So, according to the manner of quantum mechanics, an action function of DM seesaw motion should be equal to multiplication of p and λ:

$$S=m\cdot c\cdot\lambda\approx1.2\cdot10^{-39}\cdot3\cdot10^{8}\cdot1.9\cdot10^{-3}\approx6.8\cdot10^{-34}\text{J}\cdot\text{s}.$$

As it is obvious, such obtained estimation of the value of an action function of DM seesaw motion also corresponds well to the Planck`s constant h.

It is clear that obtained two estimations of S value for DM seesaw motion discover the origin of Planck`s constant h in the nature as the result of DM thermal seesaw motion at DM temperature ~2.7K. No doubt, what important to note, that if DM temperature (or any other DM parameter) will change, then it will courses the change of DM action function and, correspondingly, the change of the value of Planck`s constant h. Obviously, at that all known nature properties (of microcosm first of all) should be changed, but if all they will be changed very slowly (or in-phase to each other) then their change will not be checked by any observations.

As clear, the obtained result for h origin from DM seesaw motion gives materialistic base to the Heisenberg`s principle of uncertainty and to all other conceptions of quantum mechanics.

It should be underlined, that DM thermal seesaw motion gives full vindication of adaptability of statistical physics. Really, the DM thermal motion (at ~2.7K) means, like to known thermal Brownian motion, that the trajectories of all material bodies are get entangled in some extend and, correspondingly, they may be intersected by itself. Such real property of body trajectory rejects the one of the main postulate of classical mechanics (about nonintersecting body trajectories) and prove the truth of Boltzmann`s approach in statistical physics.

4. Estimation of Dark Matter Density in the Universe

As shown, damping length of seesaw motion of DM effective mass is equal to $\lambda\approx1.9\cdot10^{-3}$m, what taking into account the effective mass value ($m\approx1.2\cdot10^{-39}$kg) let to obtain an estimation of very interesting cosmological parameter, namely, the density of DM (ρ). The knowledge of ρ value is necessary for solving one of the important cosmological problems, which concerns the understanding of stability of our Universe.

The proposed estimation of ρ is based on reasonable assumption that damping length ($\sim 1.9 \cdot 10^{-3}$m) of DM seesaw motion includes not one but many (>10) effective masses of DM. Such assumption corresponds to known damping behavior of Langmuir longitudinal waves in plasma, where their damping length exceeds even the value of Debye screening length, which includes very many particles of plasma. For example, in the case of interslellar space plasma Debye screening length includes more than 100 active particles of plasma. The use of analogy between the damping length of DM seesaw motion and damping behavior of Langmuir longitudinal waves in plasma is reasonable as sustained CMBR producing by DM seesaw motion is similar to existing sustained waves in plasma.

As reasonable, it may be assumed that damping length of DM seesaw motion ($\sim 1.9 \cdot 10^{-3}$m) includes about 30 effective mass $m \approx 1.2 \cdot 10^{-39}$kg, what gives for the density of DM the value

$$\rho \approx m \cdot (30/\lambda)^3 \approx 5 \cdot 10^{-27} \text{kg/m}^3.$$

The ρ value, obtained at such assumption, responds well to the critical density ($\rho_c \approx 0.9 \cdot 10^{-28}$kg/m^3) of space matter for the case of stationary Universe. These results are in agreement with recent data from space vehicles (Planck and others satellite data), which have established the stable state of our Universe. Such agreement is important since all these data indicate in general the absence of the Universe inflation problem, what leads to another problem presumption of mysterious dark energy, which has no its clear physical sense yet. It may be shown, that these two problem ideas are eliminated by accounting the existence of DM, i.e. by the fact that the space is really material medium.

As known the problem assumption that our Universe is expanding was caused by observed red shift of frequency (reduction of energy) for electromagnetic waves coming from deep cosmic space. Moreover, this red shift (and reduction of wave energy) increases with increasing space distance, from which electromagnetic waves are coming. Such red shift (and reduction of energy) may be simply explained by natural dissipation of energy of electromagnetic waves while they are propagating through the filled by DM space, which is real material medium. As clear, such dissipation must increase with increasing space distance, what logically explains the observed red shift increase with space distance. This materialistic explanation of observed red shift, known as concept of tired light, is natural and evidently true since it eliminates both obviously mysterious ideas about Universe inflation, induced by physically queer assumption of Big Bang, and about physically unexplained reason of dark energy.

5. Creation of Electron-Positron Pairs from Dark Matter: on Substantiation of Positron Nature of Mysterious Ball Lightning

The existence of DM in the space and reasonable comprehension that electron-positron pairs are created from this DM (as material vortexes according to their spins) let substantiate positron nature of yet misunderstood atmospheric phenomenon – ball lightning (BL). It is shown [11, 12], that understanding of BL physical nature as a bunch of big amount of positrons let explain all observed specific properties of BL. Main known properties of BL (ball-shaped luminescence, visible size and time of life) are determined by behavior in the air of X-ray radiation from annihilation of BL positrons with electrons of the air molecules. Efficient creation of positrons in thunderclouds at streak lightning was revealed recently (2009) with the use of space telescope Fermi (by registration the X-ray radiation, which is specific for electron-positron annihilation) [13]. This fact is confirmed by mountain station observations [14], which also show the appearance at mighty streak lightning of many very hot electrons and gamma quanta. Their appearance is due to high electric fields in thunderclouds (up to 200 MV), what at mighty streak lightning leads to strong heating of electrons and to their multiplication by ionization of air molecules. As known such hot electrons at their retardation in the electric field of nucleus (of air molecules) effectively produce the bremsstrahlung (gamma quanta with MeV energy), and these gamma quanta create (also effectively in the nucleus electric field of air molecules) the electron-positron pairs.

The possibility of positron confluence (at first sight unbelievable for the same charges) may be reasonably substantiated for the case, when positrons are created concurrently and next to from all ambient DM [11, 12]. Since every positron is material vortex and its electric field results from DM polarization by its rotation, it is clear that, if all nearest DM will be consumed for creation of many positrons, so at this moment around thus created positrons will be not enough DM to produce their electric fields, which lead to their mutual repulsion. Thus, such created positrons, being an identical material vortex, may confluence in one rotating bunch. According to the energy conservation law the rotation speed of growing positron bunch should diminish so that its surface speed remains constant [12]. So electric field of positron bunch (due to DM polarization by BL rotation) should stay small and positive, as it is marked for BL in [15], where the most full data of observed BL properties are presented. It is necessary to note, that there are many evidences of witnesses, who observed BL rotation, when BL was moving near them. As to this intrinsic BL property, most indicative is the unique case, when the rotation of emerging and growing BL (after nearby impact of streak lightning into the ground) was observed and described by witness in [16].

Revealed (at explanation of positron nature of BL) conception of physical essence of electric field (and electric charge) of positron means that observed "electric charge" of any elemental particle is the result of its "electric field", produced by nearest DM "polarization" due to particle rotation [17]. Besides, this conception shows that known universal constant "electric charge" of stable elemental particle (electron, proton and their antiparticles) is the consequence of angular momentum conservation law for such rotating particle, created from DM as stable material vortex.

6. Conclusions

Reasonable materialistic assumption that observed CMBR is produced by thermal motion of DM, which fills ambient space, let substantiate, using observed properties of CMBR, both the classical physics conservation laws and also the materialistic origin of all quantum mechanics principles since DM seesaw thermal motion creates the mechanical action equal to known value of Planck's constant.

As shown, namely observed high spatial homogeneity and isotropy of CMBR and, naturally, the same properties of its producing DM, really validate the classical physics conservation laws of momentum and angular momentum, correspondingly. Whereas the conservation law of energy follows from observed stability of behavior (independence on starting moment) of physical processes in the nature.

As also shown, the Planck`s constant may be directly obtained from the features of CMBR, which is produced by equilibrium seesaw motion of DM at temperature ~2.7K, taking into account that Planck`s constant is in their essence the mechanical action too. That result obviously gives materialistic ground for Heisenberg`s principle of uncertainty and, correspondingly, for all conceptions of quantum mechanics.

As about to the applicability bounds of classical physics fundamental laws, it is clear that conservation laws for momentum and angular momentum will take place everywhere, where DM is present with its high homogeneity and isotropy (at large scale). If it is not so, as it is near massive or charged bodies, the observed deviations from this laws are taken into account as usual by using corresponding potentials (gravitational or electrical). As regard to energy conservation law, it will be applied if parameters of DM seesaw motion (or that of CMBR) will be constant or vary enough slowly during considered time.

It is necessary underline that observed (from the Earth) weak (~0.1%) dipole anisotropy of CMBR obviously indicate the existence in space of some material substance (just DM), through which the Solar system moves (at the rate ~400km/s) in direction of the Leo constellation. This feature of CMBR is greatly important since it directly demonstrates that so called "vacuum" is not empty, but it is filled by now actively sought-for DM (or "ether" as it was earlier named).

Understanding of existence of DM with observed properties of CMBR let give materialistic substantiation to conservation laws of classical physics and to principles of quantum mechanics as creates the value of Planck's constant from DM thermal motion. Such understanding also let substantiate positron nature of BL, what allow first explain all observed BL properties, and reveal physical essence of electric field (consequently, electric charge) of elemental particles, created from DM.

References

[1] Boriev I A 2014 Fundamental laws of classical and quantum physics follow from the features of microwave background radiation produced by dark matter seesaw motion *Int. J. Astron. Astrophys. and Space Science* **2** pp 7-11 http://www.openscienceonline.com/journal/archive2?journalId=703&paperId=1328

[2] Boriev I A 2015 Real state of the physical properties of space and time *International Scientific Conference "Physical Interpretations of Relativity Theory"* June 29–July 2 2015 Moscow BMSTU http://www.pirt.info/scopus/all-issues/2015/articles/Boriev.pdf

[3] http://map.gsfc.nasa.gov/universe/bb_tests_cmb.html

[4] Paolo L 2012 Open problems in particle astrophysics *Nuclear Insruments and Methods in Physics Research A* **692** pp 106-119 http://dx.doi.org/10.1016/j.nima.2012.02.002

[5] Vanderburgh W L 2014 *Studies in History and Philosophy of Modern Physics* 47 pp 62-67 http://dx.doi.org/10.1016/j.shpsb.2014.05.008

[6] Chakdar S et al 2014 Parallel universe, dark matter and invisible Higgs decays *Physics Letters B* **732** pp 343-348 http://dx.doi.org/10.1016/j.physletb.2014.03.024

[7] Davini S Enter the darkside, *Nuclear Insruments and Methods in Physics Research A* **742** pp 183-186 http://dx.doi.org/10.1016/j.nima.2013.10.066

[8] Noether E 1918 *"Invariante Varlationsprobleme"*, Nachr. d. König. Gesellsch. d. Wiss. zu Göttingen, Math-phys. Klasse (1918) pp 235-257; English translation M. A. Travel, Transport Theory and Statistical Physics 1(3) 1971,183-207.

[9] http://www.physics.ucla.edu/~cwp/articles/noether.asg/noether.html

[10] Bohm D 1952 Quntum theory *New York, Prentice-Hall, Inc.*p 728

[11] Boriev I A 2013 On the properties of electrical phenomena in the earth's atmosphere: the explanation of the physical nature of ball lightning *X international conference "Wave electrodynamics of a conducting fluid. DPO and little known forms of natural electrical discharges in the atmosphere* Yaroslavl, Yaroslavl State University July 04-08 2013 Conference proceedings pp 14-18 (in Russian) http://sites.google.com/site/blehdconference2013

[12] Boriev I A 2015 Electron-positron pair creation from dark matter: substantiation of positron nature of ball lightning *International Journal of Astronomy, Astrophysics and Space Science* **2** pp 45-50 http://www.openscienceonline.com/journal/archive2?journalId=703&paperId=2565

[13] The NASA's Fermi telescope detects antimatter emitted into space from thunderclouds www.astrogorizont.com/content/read-Fermi_ficsiruet_antimateriyu

[14] Chilingarian A, Mailyan B, Vanyan L, 2012 Recovering of the energy spectra of electrons and gamma rays coming from thunderclouds *Atmospheric Research* **114-115** pp 1-16

[15] Stakhanov I P 1996 On the physical nature of ball lightning Moscow: Scientific world p 264 (in Russian).

[16] Tar D 2006 Observation of Lightning Ball (Ball Lightning): A new phenomenological description of the phenomenon *Proceedings of the 9th International Symposium on Ball lightning (ISBL-06)* August 16-19 2006 Eindhoven The Netherlands Eds. G.C. Dijkhuis D.K.Callebaut and M.Lu pp 222-232 http://arxiv.org/ftp/arxiv/papers/0910/0910.0783.pdf

[17] Boriev I A 2016 Existence of Dark Matter and its Meaning for the Physical Essence of "Electric Field" and "Electric Charge" of the Particles created from this Matter *International Journal of Astronomy, Astrophysics and Space Science* **3** pp 1-7 http://www.openscienceonline.com/journal/archive2?journalId=703&paperId=3539

Theory of liquid crystal orientation under action of light wave field and aligning surfaces

A K Dadivanyan [1], D N Chausov[1], V V Belyaev[1,2], N N Barabanova[1], O V Chausova[3], Yu D Kuleshova[1]

[1]Moscow Region State University, 10a Radio st., 105005 Moscow, Russia
[2]RUDN University (Peoples' Friendship University of Russia), Moscow, Russia
[3]University of Technology, 42 Gagarina st., 141070, Korolev, Russia

E-mail: d.chausov@yandex.ru

Abstract. Theoretical models developed in the MRSU group under leadership of Professor Artem Dadivanyan in area of the LC orientation and photo-induced effects are presented. Angular distribution functions of the dye and liquid crystal molecules under action of intensive light beam have been derived. The number of molecules in cluster is estimated. A model of dimers formation in the photoalignment dye is suggested that explains influence of the dye molecular structure on both polar and azimuthal anchoring energy.

1. Introduction

In this paper we review main results obtained in the MRSU group under leadership of Professor Artem Dadivanyan in area of the LC orientation and photo–induced effects:

- The effect of photoalignment of liquid crystals under the action of polarized light. This technology is applied for the alignment of liquid crystals in liquid crystal displays, lenses, optical films [1–4].
- Photoinduced effects in layers of the LC, dyes and polymers [5,6].
- Investigation of properties of liquid crystal nano–colloids [7–9].

2. Liquid crystal alignment parameters

In [10] this model was expanded for the total orientation angular range from 0 to 180°.

A formula (1) for the energy of the mesogens intermolecular interaction vs. molecules orientation angle as well as order parameter S is as follows:

$$E = 2\sum_{i,j,n}\left(\mathrm{A}\left(\frac{A_{i,j,n}(S)\cos^2\alpha + b(i,j,n)}{C_{i,j,n}(S)\cos^6\alpha + F_{i,j,n}(S)\cos^4\alpha + G_{i,j,n}(S)\cos^2\alpha + h(i,j,n)} \right)^2 - \mathrm{B}\left(\frac{A_{i,j,n}(S)\cos^2\alpha + b(i,j,n)}{C_{i,j,n}(S)\cos^6\alpha + F_{i,j,n}(S)\cos^4\alpha + G_{i,j,n}(S)\cos^2\alpha + h(i,j,n)} \right)\right), \quad (1)$$

where

$$A_{i,j,n}(S)=\sum_{k=0}^{2}a_k(i,j,n)\cdot S^k\,,\ G_{i,j,n}(S)=\sum_{k=0}^{2}g_k(i,j,n)\cdot S^k\,,F_{i,j,n}(S)=\sum_{k=0}^{4}f_k(i,j,n)\cdot S^k\,,$$

$$C_{i,j,n}(S)=\sum_{k=0}^{6}c_k(i,j,n)\cdot S^k\,,$$

$a_k(i,j,n), b_k(i,j,n), c_k(i,j,n), f_k(i,j,n), g_k(i,j,n), h_k(i,j,n)$ are functions that characterize number of the interaction contacts between molecules of both upper and lower layers.

The dependence of the energy of the intermolecular interaction vs. molecules orientation angle at different values of the order parameter S is presented in Fig. 1.

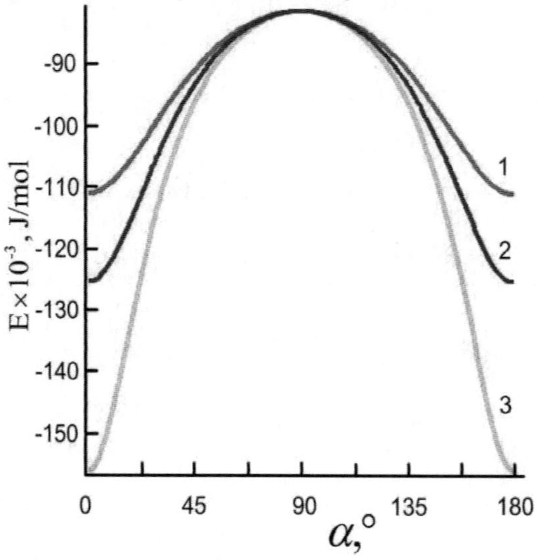

Figure 1. The dependence of the energy of the intermolecular interaction vs. molecules orientation angle at different values of the order parameter 1 – S=0.7; 2 – S=0.8; 3 – S=0.9.

If the angle α is replaced for the angle θ between the LC molecule and the normal to the XOY plane and review a case of α close to 90° (or θ close to 0) then this formula can be written in a view of Rapini–Papoular potential $E=w_1^{'}+w_2^{'}\sin^2\theta$ this formula can be written in a view of well–known Rapini–Papoular potential.

3. Distribution of dye and liquid crystal molecules orientation under action of the light wave field

In [11, 12] photoalignment properties of some azo dyes have been investigated and values of both polar and azimuthal anchoring parameters (W_θ and W_φ, accordingly).

In [11] a model of dimerization of some azo dyes molecules was developed that allows explanation of different values of the ratio W_θ/W_φ vs. dye chemical structure. When the dye molecule absorbs radiation, the nonradiative energy transition from the molecule in a triplet state to the molecule in a ground state takes place due to the exchange-resonance interaction. It occurs when electronic shells of excited and not excited molecule overlap and energy and electrons are transferred mutually [13, 14]. This interaction has an impact on the molecule structure. An assumption was made in [11] that dimers of the dye molecules are formed via hydrogen bonds between carboxyl–, hydroxyl– and CO groups, and the dimer formed has a nearly flat structure.

In [11] the following equation (2) for the variation of entropy has been derived:

$$\Delta S = \frac{1}{2} N_A k \ln(2N^d_{trans} N^d_{or}) - 2N_A k \ln(2N^m_{trans} N^m_{or}) = \\ = \frac{1}{2} R \left(\ln \frac{N^d_{trans}}{(N^m_{trans})^2} + \ln \frac{N^d_{or}}{(N^m_{or})^2} - \ln 2 \right) \tag{2}$$

where N_A is the Avogadro number, R is the gas constant.

The value of ΔS equal to 60 ± 10 J/mol·K, and the change of free enthalpy ΔG at the fusion temperature of CD–1≈ 600 K equal to –70 kJ/mol. These values indicate that the probability of the dye molecule to exist in the dimer form is by several factors higher than to be in the monomer form. Energy of dimerization for the dyes SD–1 and DR–23 was found using GAMESS software package [15], and equals to –107.5 kJ/mol and –49.5 kJ/mol, respectively. These values are approximately equal to the hydrogen bonds energy of the carboxylic and hydroxyl groups respectively. Calculated ΔS and ΔE allowed us to find the value of ΔG = –64 kJ/mol and –24 kJ/mol for SD–1 and DR–23 respectively. Consequently, these dyes must be in the dimer form, too.

For the dyes BY and CBY ΔE = –25 kJ/mol, that results in ΔG = +11 kJ/mol and the probability of forming the dimer is less than 10^{-3}.

Anchoring energy is proportional to the torque resistant to the molecule's rotation. This torque is proportional to the particle (dimer or molecule) cross-section in the corresponding plane. Since the molecule orientation is planar, W_θ is proportional to the dimer cross–section σ_θ in the plane intersecting the easy rotation axes of both molecules, while W_φ is proportional to the section σ_φ in the plane orthogonal to easy rotation axes.

In [12] values of a $\cos^2\varphi$ parameter, which characterizes orientation order of both molecules and dimers, were obtained as well as energy of their orientation process. It is of order of $7\cdot10^{-34}$ J per a molecule. Number of cooperatively oriented dye molecules was estimated. Therefore the number of cooperatively oriented molecules is of the order of 10^{13} at least.

The angles distribution functions for different dimers were found. The distribution function of the dimer in the electric field of the angle φ':

$$\rho(\varphi') = C \cdot e^{-\frac{NU(\varphi')}{kT}}, \text{ where } C = \frac{1}{\int_0^{2\pi} e^{-\frac{NU(\varphi')}{kT}} d\varphi'}.$$

Dipole electrical transition moments directed on angles close to 90º to the direction of maximal polarizability of dimers were revealed.

The transition dipole moments directions almost coincide with the direction of the highest polarizability of the molecule, while in the dimers one of the transition moments is directed nearly perpendicular. Consequently, when the dimers are oriented, certain transition dipole moments will form angles close to 90º with the polarization plane of the orienting radiation.

A consequence of the absorption mechanism of the photoalignment is that the photon absorption can result in dissociation of the hydrogen bonds between the molecules. For the CD–1 dye the dimerization energy is equal to $1.2\cdot10^{-19}$ J (72 kJ/mol) that is of the same order as the energy of the photons used for the photoalignment (for the blue light with λ=450 nm the photons energy is $4.4\cdot10^{-19}$ J (266 kJ/mol)). This fact must be taken into account considering the absorption mechanism of the photoalignment.

4. Conclusion

At the first time there were derived angular distribution functions of the dye and liquid crystal molecules under action of intensive light beam. This model results in values of the interaction energy of the molecule with the light wave electric field as well as molecules number in cluster. Both these directions are integrated with a model of dimers formation in a photoalignment dye that explains influence of the dye molecular structure on both polar and azimuthal anchoring energy.

5. Acknowledgement
The paper is prepared under partial support of Russian Foundation for Basic Researches (RFBR) (project No. 17-47-500752 and 17-47-500757 with Moscow Region Government).

References
[1] Chigrinov V G., Kozenkov V M, Novoseletsky N V., Reshetnyak V Y., Reznikov Y A., Schadt M., Schmitt K., Process for making photopolymers having varying molecular orientation using light to orient and polymerize. US Patent No. 5389698 Publication date 1995/2/14. Application number 08/125,006.
[2] Dyadyusha A G, Kozenkov V M, Marusiy T Y, Reznikov Yu A, Reshetnyak V Yu, Khizhnyak A I, *Ukrainskii Fizicheskii Zhurnal.* 1991. **36**. P.1059.
[3] Chigrinov V G, Kozenkov V M, Kwok H, Photoalignment of liquid crystals: physics and applications, **36**, Wiley, 2008.
[4] Yaroshchuk O, Reznikov Y, *J Mater Chem.* 2012. **22**. P.286.
[5] Slussarenko S, Francescangeli O, Simoni F, Reznikov Y, *Applied physics letters* 1997. **71(25)**. P.3613.
[6] Boichuk V, Kucheev S, Parka J, Reshetnyak V, Reznikov Y, Shiyanovskaya I, Singer K D, Slussarenko S, *Journal of Applied Physics* 2001. **90**(12). P.5963.
[7] Francescangeli, Slussarenko S, Simoni F, Andrienko D, Reshetnyak V, Reznikov Yu *Physical Review Letters* 1999. **82**(9). 1855.
[8] Reznikov Y, Glushchenko A, Garbovskiy Y. *Liquid crystals with nano and microparticles (in 2 volumes).* Singapore: World Scientific Publishing Co. Pte. Ltd.; 2017. P.657. Reznikov Y, Buchnev O, Tereshchenko O, et al. *App Phys Lett.* 2003. **82**(12). P.1917.
[9] Chausov D N, Dadivanyan A K, Belyaev V V, Noah O V, *Molecular Crystals and Liquid Crystals.* 2015. **611**. 1. P.21.
[10] Barabanova N N, Belyaev V V, Bogdanov D L, Chigrinov V G, Dadivanyan A K, Nazarov A P, Noah O V, *Journal of the Society for Information Display*. 2015. **23.** 10. P.486.
[11] Dadivanyan A K, Nazarov A P, Barabanova N N, Belyaev V V, Bogdanov D L, Borisenko O V, *EuroDisplay conference*, Berlin, Germany, from 31 October to 2 November 2017. Abstracts. P–06.
[12] Th. Förster *Ann. d. Physik* 1948. **2**. P.55.
[13] Terenin A N, *Uspehi fiz. nauk*, 1956. **58**. P.37.
[14] Palermo V, Biscarini F, Zannoni C, *Phys. Rev. E.* 1998. **57**. P.2519.

IOP Conf. Series: Journal of Physics: Conf. Series **996** (2018) 012019 doi:10.1088/1742-6596/996/1/012019

Interaction of dyes CD–1 and SD–1 with the surface of oligodimethysiloxane

D N Chausov

Moscow Region State University (MRSU), Moscow, Russia

E-mail: d.chausov@yandex.ru

Abstract. We carried out the modeling orientation of the dyes CD–1 and SD–1 relative to the surface of oligodimethysiloxane using the atom–atom potentials method. We have discovered the dependence of the interaction energy in dyes molecules on the angles which characterizes their orientation relative to the surface of the oligodimethysiloxane crystal. It was found out that the obtained energy value of interaction with the surface can explain weak adhesive qualities of the dyes and the orientation type relative to the surface. We identified the break–loose force for the dyes on the oligodimethysiloxane crystal surface.

1. Introduction

Nowadays the most prospective materials for LC displays are so–called photo-anisotropic materials oriented on the surface of different materials which have organic and non–organic origin. They are used or potentially useful for polarization-optical structures as thin–film anisotropic-optical protective and identification equipment for various industrial products.

The orientation of aliphatic hydrocarbons absorbed on the crystal surfaces is well studied. [1–6]. At the same time, there are few studies on dyes orientation on the crystal surface.

The goal of the study is to study intermolecular interactions of the dyes CD–1 and SD–1 with the oligodimethysiloxane surface. Many derivates from oligodimethysiloxane or silicones are known as the materials that orientate liquid crystals [7–13].

2. Simulation Method

Structural formulae of photochemically stable materials SD–1 and CD–1 are shown in Fig. 1, the molecules lined up according to the coordinates of their atoms. In the B dyes molecules the surface of phenyl rings of carboxyl groups and –N=N– groups make up an angle of 16° between two phenyl groups in biphenyl. Atoms coordinate optimization at lining up dyes was carried out using the special software for physics and chemistry modeling GROMACS.

IOP Conf. Series: Journal of Physics: Conf. Series **996** (2018) 012019 doi:10.1088/1742-6596/996/1/012019

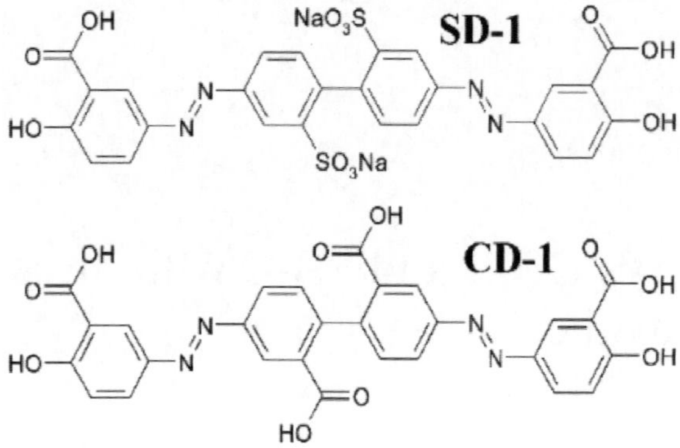

Figure 1. Dyes SD–1 and CD–1 structural formula

Within the structure of oligodimethysiloxane silicon and oxygen atoms form a zigzag–like chain. The distance between atoms Si–O in the chain is 63Å, the distance Si–C – 1.88Å. Valence angle value Si–O–Si is 168°, the angle O–Si–O – 109.5° fig. 2.

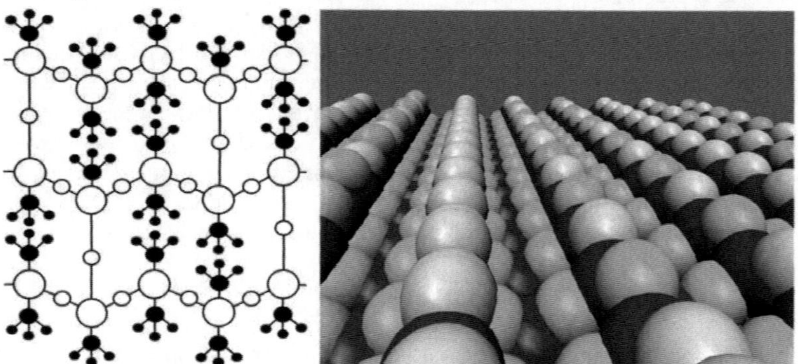

Figure 2. Spatial molecules structure: ◯ – (Si) silicon, ○ – (O) oxygen, • – (C) carbon (a); oligodimethysiloxane surface 3D model (b)

Orientation of the molecule to the crystal surface is defined by angles θ, φ, ψ (fig. 3) which changed from 0° to 180° at a pitch of 10°. Here X,Y,Z – is the system of axes connected with the crystal surface; 1,2,3 – is the system of axes connected with the macromolecule; θ – the polar angle of axis 1 in rigid axes system; φ – azimuth of axis 1 (the angle of XZ with 1Z surfaces); ψ – azimuth of the moving coordinates system (the angle of 1Z surface with 12 surface).

The initial distance value *a* from a molecule centre of attraction to the surface going through the centers of attraction of oligodimethysiloxane atoms, which form the surface, was taken as 3 Å and minimizing the energy was performed according to altitude *a*.

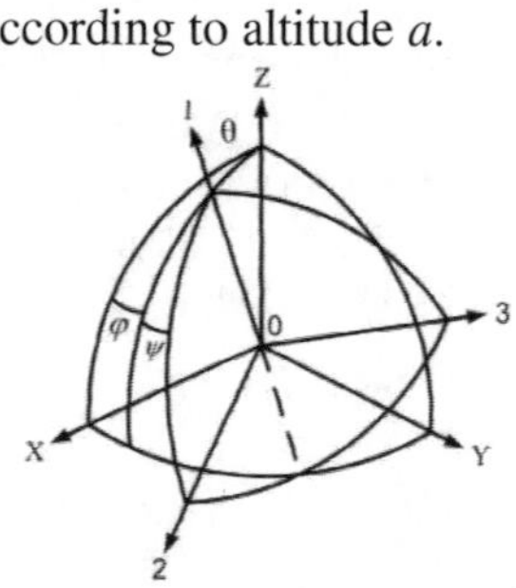

Figure 3. Mutual arrangement of the coordinates systems, connected with the molecule (1, 2, 3) and the surface (x, y, z). θ – the polar angle of axis 1 in the system x, y, z, φ – the azimuth of axis 1 (the angle of the surface xz with 1z).

The calculations were carried out using atom–atom potentials method [14–17]. The interaction energy E is shown as a sum of energy of interaction values between the atoms of the molecule and the atoms of the crystal surface, which is determined by the correlation (Lennard–Jones potential) [17]:

$$E = \sum_{ij} E_{ij},$$

$$E_{ij} = 4\varepsilon\left[\left(\frac{\sigma}{r_{ij}}\right)^{12} - \left(\frac{\sigma}{r_{ij}}\right)^{6}\right] \quad (1)$$

in which $\varepsilon > 0$ – depth parameter, σ – distance between certain atoms, at which $E_{ij}=0$; r_{ij} – the distance between the atom surface i and the atom of the molecule j.

The interaction energy was found by the correlation (1), where the interaction parameters values for groups C'...C', Si...Si, O...O were taken from the works [24–26]. Interaction potentials C'...O, and C'...Si were calculated using the well–proven additivity concept of van der Waals radiuses and the rule of geometric mean which were found in [27]. We can mention that using these values of atom–atom potentials we obtained the values for molecules orientation for liquid crystal octylcyandiphenyl on the surface of graphite and polyethylene crystals [20–21] which are in the accordance with the test data [22–23].

3. Results of numerical simulation

As the result we proved the dependence of the interaction energy of the oligodimethysiloxane crystal and dyes molecules SD–1 и CD–1 on the orientation angles θ, φ, ψ.

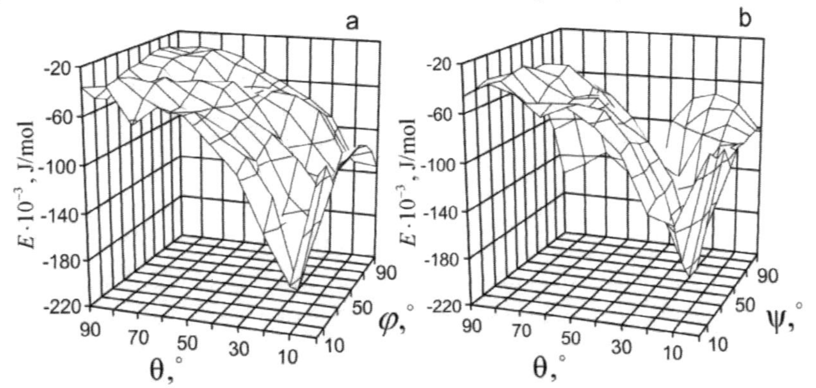

Figure 4. Dependence of the interaction energy in the oligodimethysiloxane system – CD–1 on orientation angles θ, φ, at ψ=30° (a) and orientation angles θ, ψ, at φ =10° (b)

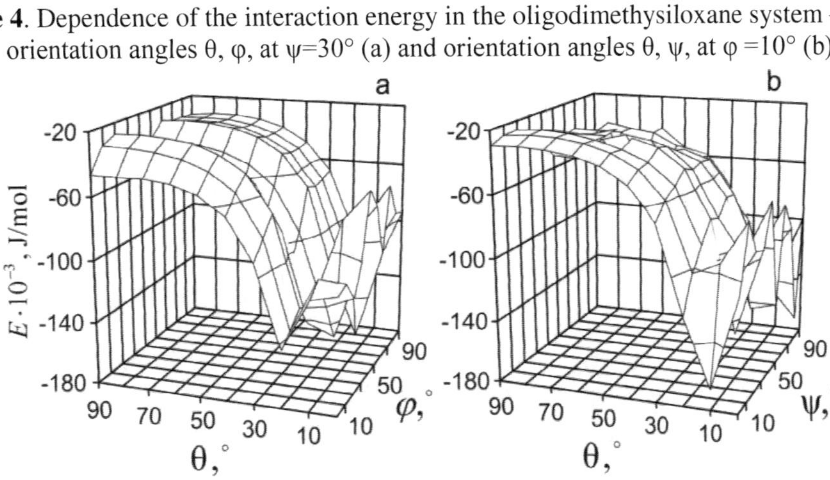

Figure 5. Dependence of the interaction energy in the oligodimethysiloxane system – SD–1 on orientation angles θ, φ, at ψ=30° (a) and orientation angles θ, ψ, at φ =10° (b)

We can see in figures 4 – 5, that the intermolecular interaction energy of the dyes on the oligodimethysiloxane crystal surface essentially depends on three orientation angles θ, φ, ψ, herewith, there is an area of orientation state for the dyes molecules, whose interaction energy of the oligodimethysiloxane surface is massively more than the energy in other state, (Table 1).

Table 1. Interaction energy of dye molecules with oligodimethysiloxane surface

Dyes	θ,°	φ,°	ψ,°	E_{max}, kJ/mol	θ,°	φ,°	ψ,°	E_{min}, kJ/mol
CD–1	90	40	40	–27.75	10	10	30	–191.31
SD–1	80	80	30	–23.24	10	10	30	–167.25

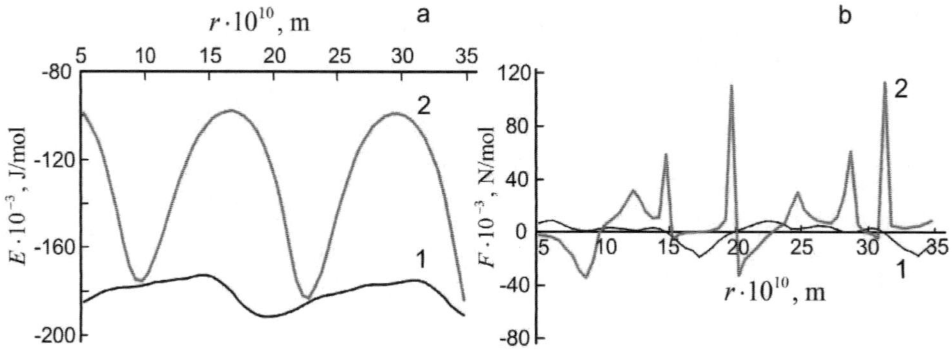

Figure 6. Energy a) and force b) of the interaction between CD–1 at the main line symmetry OX direction shift 1) and OY direction 2) of oligodimethysiloxane, dye orientation angles are θ=10°, φ =10°, ψ=30°

In figure 6 a) curve 1 shows that the anisotropy of interaction energy for molecule CD–1 at the direction shift along silicon–oxygen atomic bridge Si–O–Si in oligodimethysiloxane is not more than 5%, herewith the force $F = \partial E_i / \partial x_i$ (fig. 6 b curve 1) is near zero. At the same time, curve 2 (fig. 6 a) shows that anisotropy of the interaction energy for molecule CD–1 at a perpendicular shift to chain Si–O is ~50%. Due to it we can evaluate the orientation degree of the dye in different directions in oligodimethysiloxane.

4. Acknowledgement

The paper is prepared under partial support of Russian Foundation for Basic Researches (RFBR) (project No. №18–07–00727_a) and Presidential Grant No. МК–3120.2018.9.

References

[1] Hentschke R, Schurmann B L, *Surf. Sci.* 1992. **262**. P.180.
[2] Israelachvili J N, Intermolecular and Surface Forces (Academic, London, 1992).
[3] Foster J S, Frommer J E, *Nature* (London), 1988. **333**. P.542.
[4] Hara M , et al. *Nature* (London), 1990. **344**. P.228.
[5] Poulin J C , Kagan H B, *Acad. Sci., Ser.II: Mec., Phys., Chim., Sci. Terre Univers*, 1991. **313**. P.1533.
[6] Dadivanyan A K., Kirakosyan L H, Ignatov Y A, *Bulletin of the Moscow State Regional University Series «Physics and Mathematics»*, 2005. **1**. P.59.
[7] Cognard, J. Alignment of Nematic Liquid Crystals and Their Mixtures, *Molecular Crystals and Liquid Crystals*, Suppl. 1 (Gordon and Breach, London, 1982).
[8] Negoro M, Kawata K. "Process for vertical alignment of rod–like liquid crystal molecules", US Patent 6,726,966, Publication Date: April 27, 2004.

[9] Belyaev V V, Mazaeva V G, Minko A A, Timofeev S N, *Bulletin of the Moscow State Regional University Series «Physics and Mathematics»*, 2010. **1**. P.52.
[10] Sobolevsky M V, Mazaeva V G, Kovalenko V I, Belyaev V V, Kalashnikov A Y, *Mol. Cryst. & Liq. Cryst.*, 1999. **329**. P.293.
[11] Mazaeva V G, Belyaev V V, Sobolevsky M V, *J. SID*. 2005. **13**. P.373
[12] Muravsky A A, Muravsky A A, Mazaeva V G, Belyaev V V, *J. SID*, 2005. **13**. P.349.
[13] Storozhenko P A, Mazaeva V G, Lotarev M B, Nazarova D V, Polivanov A H, Belyaev V V, Patent RF№2283338. Inventions bulletin on the RF Issue 25, 10.09.2006, prior. 07.07.2005.
[14] Pashinina Y M, Chausov D N, Dadivanyan A K, *Bulletin of the Moscow State Regional University Series «Physics and Mathematics»*, 2010. **3**. P.48.
[15] Dadivanyan A K, Noa O V, Chausov D N, *Bulletin of the Moscow State Regional University Series «Physics and Mathematics»*, 2007. **1**. P.96.
[16] Dadivanyan A K, Chausov D N, *Bulletin of the Moscow State Regional University Series «Physics and Mathematics»*, 2007. **2**. P.47.
[17] Jones J E, *Proc. R. Soc. Lond.* 1924. **106**. P.463.
[18] Glushko N K, Gubina L N, Dubrovitsky V I, Smekhov F M, Kabanov N M, Dadivanyan A K, *Macromolecular compounds A*, 1989. **31**. P.2224.
[19] Katz L, Levinthal S, *Ann.Rev. of Biophys and Bioeng*. 1972. **1**. P.465.
[20] Dadivanyan A K, Chausov D N, Pashinina Y M, Belyaev V V, *Zhidk. krist. ikh prakt. ispol'z = Liq. Cryst. and their Appl.* 2010. **4(34)**. P.61.
[21] Dadivanyan A K, Pashinina Y M, Chausov D N, Belyaev V V, and Solomatin A S, *Mol. Cryst. & Liq. Cryst.*, 2011. **545**. P.159/[1383].
[22] Poulin J-C, Kagan H B, Acad C R, *Sci., Ser. II: Mec., Phys., Chim., Sci. Terre Univers*. 1991. **313**. P.1533.
[23] Walba D M, Stevens F, Parks D C, Clark N A, Wand M D, *Science*. 1995. **267**. P.1144.
[24] Dashevsky V G, Struchkov Y T, Akopyan Z A. *Mag. struct. chem.*, 1966. **7**. P.408.
[25] Ahmed N A., Kitaigorodsky A I, Mirskaya K V, *Acta Cryst.* 1971. **B.27**. P.867.
[26] Levitt M *J, Mol. Biol.* 1974. **82**. P.393.
[27] Chausov D N, Svetlov M V, Belyaev V V, Dadivanyan A K, *Bulletin of the Moscow State Regional University Series «Physics and Mathematics»* 2012. **2**. P.162.

Variational Approach in the Theory of Liquid-Crystal State

Gevorkyan E V

Moscow Region State University (MRSU), Moscow, Russia

E-mail: gevev@rambler.ru

Abstract. The variational calculus by Leonhard Euler is the basis for modern mathematics and theoretical physics. The efficiency of variational approach in statistical theory of liquid-crystal state and in general case in condensed state theory is shown. The developed approach in particular allows us to introduce correctly effective pair interactions and optimize the simple models of liquid crystals with help of realistic intermolecular potentials.

1. Introduction

The calculus of variations [1-3], first appeared in the work of Euler in 1744 [1] and described in detail in "The Elements of the Calculus of Variations" [3] in 1766, was the foundation not only of mathematics but also of theoretical physics.

Variational methods are characterized by the effectiveness, generality and universality of the approach and give a complete look to the theory.

In the molecular-statistical and thermodynamic theory these are variational approaches based on the variational principles of Gibbs W, Boltzmann L, Bogolyubov N N, Onsager L, Prigogine I and others.

Liquid crystals are characterized by the structural diversity of the mesophase and the complex structure and the interaction of structural elements (in particular, mesogenic molecules).

They are condensed molecular systems with orientational long-range order and incomplete translational order. In accordance with this "structural" definition, their distinctive feature is the strong relationship of translational and internal, orientational molecular variables. Thus, the "dimensionality" of the statistical problem is at least twice as high as that of simple systems [4,5].

2. Theory

Let H(1, ...,N) be Hamiltonian of original realistic model of liquid crystal with internal degrees of freedom (where m= ($\mathbf{r}_m$, $\mathbf{\Omega}_m$, ζ_m)) and H_0 be Hamiltonian of simplified exactly solvable "approximating" model (for example, with multiple or discrete variables, etc.).

The Bogolyubov variational principle follows from the convexity of the continuous functional

$$\mathrm{F}\left[\mathrm{H}_0+\mathrm{A}\right]=-kT\ln\operatorname{tr}\exp\left(-\left(\mathrm{H}_0+\mathrm{A}\right)/kT\right) , \tag{1}$$

where, k – Bolsmann constant, T – temperature, $A=H-H_0$

$$\delta_{\mathrm{A}}^{2}\,\mathrm{F}\left[\mathrm{H}_0+\mathrm{A}\right]=-k^{-1}T^{-1}\left\langle\delta\mathrm{A}\middle|\delta\mathrm{A}\right\rangle_{\mathrm{H}_0+\mathrm{A}}\leq 0 \tag{2}$$

Where $\langle B\rangle$ is average value and $\langle B|C\rangle = \langle BC\rangle - \langle B\rangle\langle C\rangle$ is the second semiinvariant (cumulant) over Gibbs canonical distribution.

It follows

$$F[H_0] + \langle H - H_0\rangle_H \le F[H] \le F[H_0] + \langle H - H_0\rangle_0 = F_M \ , \tag{3}$$

The universal approach to a molecular theory of liquid-crystal state is developed on the basis of this double inequality (general variational principle by Bogolubov N N) that can also be used for qualitative estimate of approximations accuracy.

Functional F_M on the right side of this inequality is the model Helmholtz free energy. Its minimum condition over approximating Hamiltonian parameters H_0 is

$$\langle H - H_0 | \delta H_0\rangle_0 = 0 \tag{4}$$

After that F_M gives the best approximation F[H], and thus determines the approximate expression for the Helmholtz energy of initial molecular model of a liquid crystal.

In a classic case we substitute Hamilton operators H, H_0 for potential energies U_N, V_0. At that F_M is the configuration part of the Helmholtz free energy.

The initial model can include translation, orientation, conformation variables and three-, four-partial interactions. In the framework of a unified variational approach it is possible to approximate this complex realistic model by much simpler "solvable" models, for example, with a spatial lattice structure, with a discrete set of possible orientations, with separation of spatial **r**, conformation ζ and orientation $\mathbf{\Omega}$ variables, with the effective pair interactions and others. Moreover, in each case we get best-in-class approximation model of Helmholtz function and thermodynamic properties of a liquid crystal. In particular, additive models

$$V_0(\mathbf{1},\dots,\mathbf{N}) = V_t(\mathbf{r}_1,\dots,\mathbf{r}_N) + V_o(\mathbf{\Omega}_1,\dots,\mathbf{\Omega}_N) + V_c(\zeta_1,\dots,\zeta_N) \tag{5}$$

or

$$V_0(\mathbf{1},\dots,\mathbf{N}) = \sum_{i=1}^{N} V_i(\mathbf{i}) \tag{6}$$

(a self-consistent field) correspond to the statistical independence (or "uncoupling") of relevant variables. But in this case, unlike the similar approximation of distribution functions multiplicativity, the received results will be thermodynamically consistent.

Many-particle interactions play an important role in condensed matter. Potential energy is

$$U_N = \sum_{p=2}^{n} \frac{1}{p!} \sum_{i_1 \ne \dots \ne i_p} \Phi_p(\mathbf{i}_1,\dots,\mathbf{i}_p) \tag{7}$$

Let us find the equations defining effective potentials of pair intermolecular interactions $\Phi_2^{\text{eff}}(\dots)$ that best take into account (n=4) the true many-particle interaction $\Phi_3(\dots)$ and $\Phi_4(\dots)$.

Equation (4) gives

$$\left\langle U_N - \sum_{i<j} \Phi_2^{\text{eff}}(\mathbf{i},\mathbf{j}) \middle| \delta\Phi_2^{\text{eff}}(\mathbf{i},\mathbf{j}) \right\rangle_{\text{eff}} = 0 \tag{8}$$

$$\Phi_2^{\text{eff}}(\mathbf{i},\mathbf{j}) = \Phi_2(\mathbf{i},\mathbf{j}) + \frac{1}{\rho_2^{\text{eff}}(\mathbf{i},\mathbf{j})}\Big(\sum_{i\ne j\ne i_3} \int \Phi_3(\mathbf{i},\mathbf{j},\mathbf{i}_3)\rho_3^{\text{eff}}(\mathbf{i},\mathbf{j},\mathbf{i}_3)d\mathbf{i}_3 + \\ + \frac{1}{2}\sum_{i\ne j\ne i_3\ne i_4} \int \Phi_4(\mathbf{i},\mathbf{j},\mathbf{i}_3,\mathbf{i}_4)\rho_4^{\text{eff}}(\mathbf{i},\mathbf{j},\mathbf{i}_3,\mathbf{i}_4)d\mathbf{i}_3 d\mathbf{i}_4\Big). \tag{9}$$

However, effective potentials of pair intermolecular interactions (9), unlike the true ones, depend on the thermodynamic state parameters and the short-range order. This makes their "calibration" complicated, especially when used in the description of phase transitions.

For simplest models with traditional expansion of potentials in a series of spherical and Fourier harmonics in self-consistent field approximation the variational approach allows us to simplify significantly a system of equations obtained. The self-consistent potential involves the contributions of high order harmonics in average form [5].

To describe a liquid-crystal state it is fundamentally important to take into account the pairwise repulsion of solid anisometric cores. The attempts to use only the anisotropic van-der-Waals attraction of molecules on a lattice (the simplest classic model of the molecular field by Mayer – Saupe) to obtain the agreement with an experiment require interaction constant to be more than two orders of magnitude greater than the real values. Variational theory allows to take into account the repulsion of solid cores by taking a system with such interaction as a base.

$$V_0(\mathbf{1},\ldots,\mathbf{N}) = \sum_{i\neq j} V_{ij}^{hc} + \sum_i V(i) \quad , \tag{10}$$

where the hard core of approximating potential V_{ij}^{hc} must match or include the core source of real interaction. Hence,

$$(V(\mathbf{1}) - \Phi(\mathbf{1}))\rho_1^0(\mathbf{1}) = \sum_{p=2}^{4} \frac{1}{(p-1)!} \sum_{i_2\neq\ldots\neq i_p\neq 1} \int \Phi_p^{soft}(\mathbf{1},\mathbf{i}_2,\ldots,\mathbf{i}_p)\rho_p^0(\mathbf{1},\mathbf{i}_2,\ldots,\mathbf{i}_p)d\mathbf{i}_3\ldots d\mathbf{i}_p, \tag{11}$$

where $\Phi_p^{soft}(\ldots)$ is the "soft" part of the intermolecular potentials without hard core. This way we can improve Onsager and van-der-Waals type formulas.

3. Conclusion

Thus, the developed approach makes it possible to optimize the simple models of liquid crystals using realistic intermolecular potentials.

The considered approach allows to take into account many-particle interactions and introduce effective pairwise intermolecular interactions.

The consistent description of liquid crystals structure involves hierarchy of spatial scales [6]. The approach may be useful for the solution of this problem.

On the whole, the variational approach can be successfully applied to the development of liquid crystals theory on the unified basis due to its universality and effectiveness.

References

[1] Euler L Methodus inveniendi lineas curvas proprietate maximi minimive gaudentes. Lausanne-Geneva, 1744.

[2] Lagrange J L Essai d'une nouvelle methode pour determiner les maxima et les minima des formules integrales indefinies // Torino, Miscellanea Taurinensia, 1762. 2. P. 335-362.

[3] Euler L Elementa calculi variationum // St. Petersburg, Novi Commentarii academiae scientiarum imperiales Petropolitanae, 1766. 10, P. 51-93.

[4] The Molecular physics of liquid crystals (Ed G R Luckhurst, G W Gray) Academic Press, 1979. 494 P.

[5] Bazarov I P , Gevorkyan E V The Statistical Physics of Liquid Crystals. Moscow University Press. 1992. 496 P. (in Russian).

[6] Gevorkyan E V Hierarchy of spatial scales and relaxation times in liquid-crystal systems. Bulletin of Moscow Region State University. Phys.Math. 2016. no. 4. P.24-31. DOI: 10.18384/2310-7251-2016-4-24-31.

IOP Conf. Series: Journal of Physics: Conf. Series **996** (2018) 012021 doi:10.1088/1742-6596/996/1/012021

Controllable light beam shaper in LED lamp

K D Nessemon[1], I V Popov[2], V V Belyaev[1,3], A A Belyaev[3], V K Velichko[3]

[1]RUDN University (Peoples' Friendship University of Russia), Moscow, Russia
[2]Kometa Corporation, Moscow, Russia
[3]Moscow Region State University, 10a Radio st., 105005 Moscow, Russia

E-mail: nessemon@yandex.ru, vic_belyaev@mail.ru

Abstract. A design of an LED lamp with controllable shape and intensity of the light beam formed is described. The effect is achieved by dividing the light irradiating area for four section and their special fixation and fastening.

1. Introduction and problem definition

LED lamps are the lighting devices of new generation having a high light efficiency, a longevity, a small energy consumption [1,2].

Problems of LED lamps are as follows: light effectiveness, heat irradiation when using at various operating conditions. One of important problems is adaptation of the lamp to a specific workplace with a possibility of a variation of the illuminating intensity on it [3,4].

There is a set of the LED lamps having the various forms and mechanisms of fastening, for example, [5]. However when using these lighting devices it is impossible to create the given shape of the light spot since there is no possibility of management of a luminous flux – its intensity and orientation. Also at the operation of the LED lighting arrays operating temperature of a crystal increases that negatively affects exploitation performances of the light-emitting diode – life time and luminous flux are decreased, the radiation spectrum and electric parameters are changed.

The purpose of this work is development of the device providing the given direction (concentration or a scattering) of the luminous flux for the necessary type of illumination, namely common uniform, common local as well as achievement of high cooling efficiency with the reliable heat-removing device.

Let's call this device as controllable light shaper (CLS). It can be used as the lighting array in the production facilities and living conditions, both in rooms, and in outdoor areas for illumination ensuring.

2. CLS design

The aggregate CLS design is presented in fig. 1 [6]. The CLS comprises four sections. Each section consists of the basis on which profile edges of a radiator, a printed circuit board with the light-emitting diodes and an optical element (fig. 1,2) are located.

Because of extrusive manufacturing techniques all metal details of the design have profile type. At these manufacturing techniques, the surface area of the radiator is most increased for decrease of its thermal resistance for the heat transfer from the radiator into a surrounding medium.

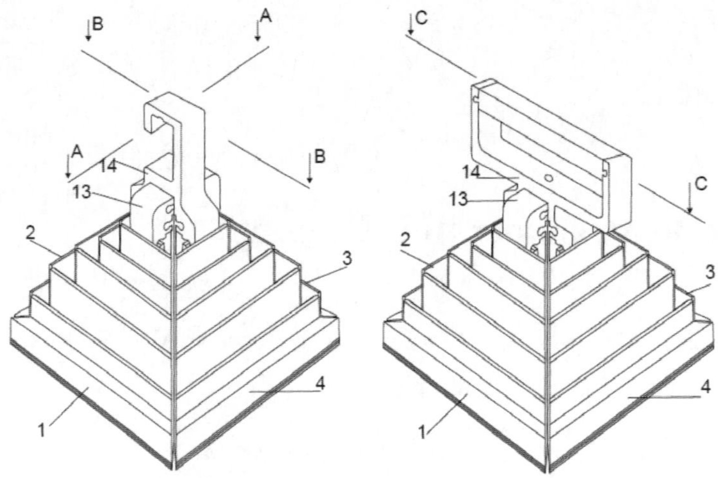

Fig.1. Axonometric drawing of the controllable light shaper (CLS). 1-4 are the CLS sections. 13 and 14 are the lower and top fixing elements for the sections 1-4 fixation and fastening.

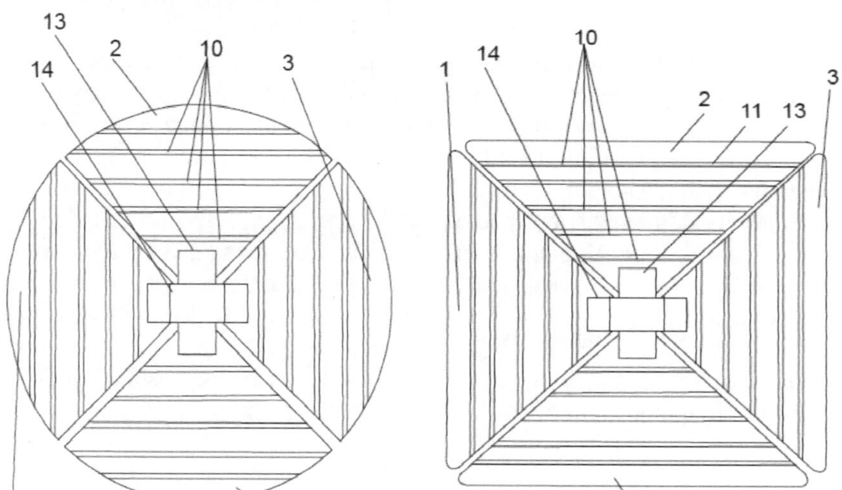

Fig. 2. The device top view in the form of a circle or rectangular in the plan. 1-4 are the sections of the controllable light concentrator. 10 are the heat-removing radiators. 13 and 14 are the lower and top fixing elements for fixing and fastening sections 1-4 among themselves. There are not shown on the drawing the elements: the basis 5, the printed circuit board 6 with light-emitting diodes, the printed circuit board 6 the optical element 7 placed on the board 6. The heat-removing radiators are a part of the heat-removing element 8 of each section located on a substrate 9.

Owing to the fixing elements which connect all four parts of the controllable light concentrator it is possible to position each part under a different corner, regulating the direction of the luminous flux that allows controlling the light spot sizes (fig. 3).

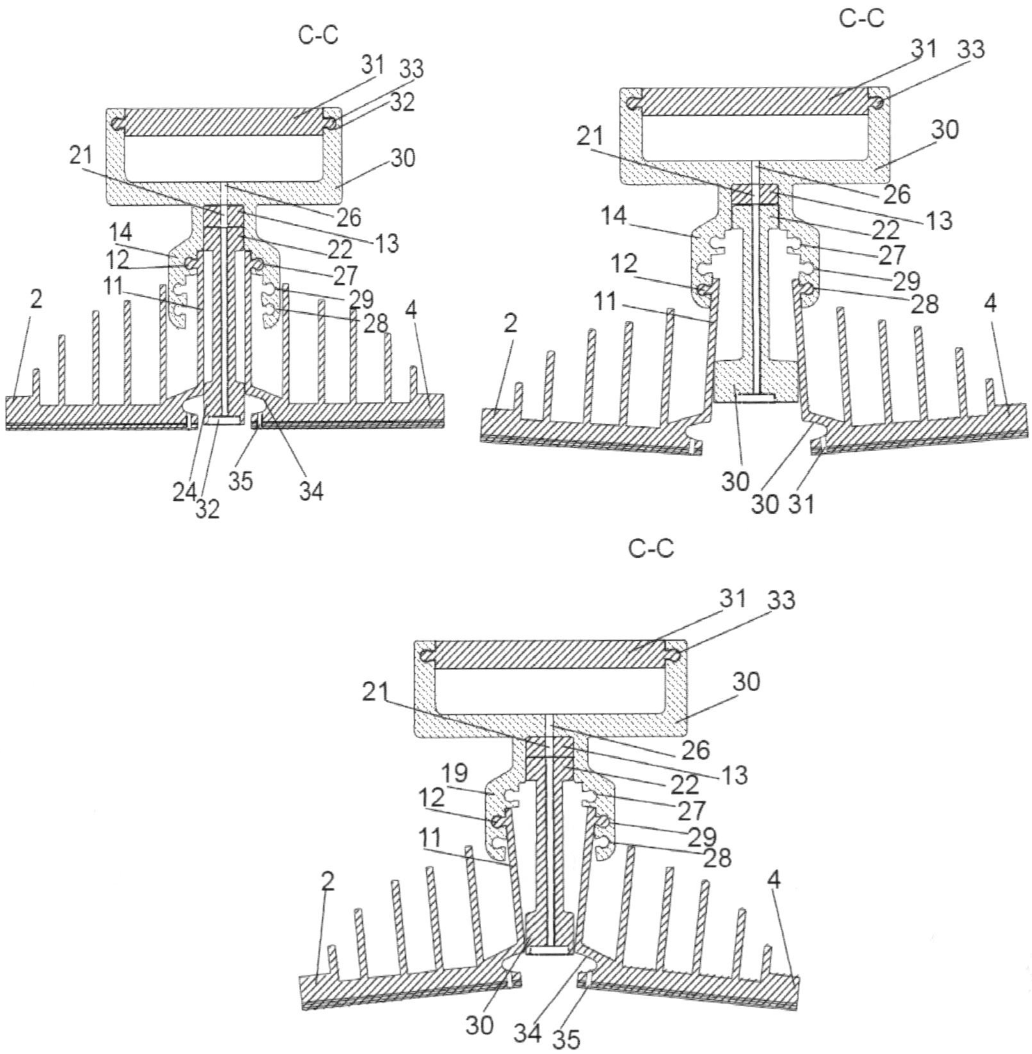

Fig. 3. The CLS front view, cross-section C-C. The arrangement of each section is horizontal, inclined up or down in relation to the central vertical axis.

As shown in fig.3 the lower and top fixing elements 13 and 14 have grooves inside, they are positioned symmetrically in relation to the central axis. Thickenings 12 of the extreme ledges of the heat-removing elements 11 are inserted into these grooves.

The CLS has a clamp 19 with a through hole which is on the same vertical axis with a through hole of the lower fixing element 21. The clamp 19 is performed rectangular in its cross-section, it has the top thickening 22 with longitudinal ledges on two opposite sides as well as the lower thickening with a rectangular aperture. The top fixing element 14 has also through hole of the top fixing element 26. All these through holes are provided for passing of wires to them from the printed circuit board for their connection with control units. The lower thickenings of the clamp 19 performed rectangular in its

IOP Conf. Series: Journal of Physics: Conf. Series **996** (2018) 012021 doi:10.1088/1742-6596/996/1/012021

section can be performed the different size longwise of the rectangle side in section depending on by what corner up or down the sections 1-4 need to be inclined, or to be arranged horizontally.

The upper part of the top fixing element can be performed in the form of a hook for hanging the device, for example, on the room ceiling. It can be also executed in the form of an inverted letter П for the placement of the control systems of the printed circuit boards in its space. In such version a cover 31 inserted into grooves 32 of the ledges 33 is used too.

Heat-removing radiators 10 can be performed as profiled to decrease the thermal resistance. For formation of the convection fluxes they settle down apart from each other. They have different height. At the same time the radiators 10 have larger height on the one side the heat-removing device, than on another side.

At the horizontal arrangement of the sections one obtains the unidirectional light beam, at an arrangement of the sections at an angle up one obtains a scattered luminescence and at an arrangement of the sections at an angle down one obtains a concentrated luminous flux with higher intensity. In this case the area of the light spot is less. It is illustrated by calculations of the illuminating intensity created on platforms under lamps of different design (Table 1). We suppose that each section with the size d has the uniform brightness distribution. The illuminating intensity with Gaussian distribution on the coordinates is created on the platform under the lamp. At the horizontal arrangement of the lamp sections the light spot has the size of 2.8 d on the level 0.5 from its maximum. At the inclined down arrangement of the lamp sections the light spot has the size of 1.5d. The illuminating intensity in the spot center of is increase by 1.5 times. At the inclined up arrangement of the lamp sections the light spot has the size of 3.5d. The illuminating intensity decreases in its maximum level by only 20%. The design of both fixing elements and grooves in them provides an inclination of the sections 1-4 from 3° to 15°.

Table 1. Illumination generated on platforms under the CLS with sections that have size d.

Coordinate, x/(d/2)	Relative illumination I/I_0			
	From one CLS	From two CLS with the horizontal sections location	From two CLS with the inclined down sections location	From two CLS with the inclined up sections location
0	1	1.34	2	0.66
1	0.67	1.33	1.34	0.82
2	0.33	0.82	0.67	1.08
3	0.15	0.41	0.30	0.69
4	0.08	0.17	0.15	0.33

If one pair of the lamp sections is inclined down, and another one is inclined up it is possible to create the light spot of the elongated form.

3. Conclusions

The device providing the preassigned direction of a luminous flux (concentration or scattering) for necessary type of the illumination (uniform or localized) is developed. It is enough for variation over a wide range of illuminating intensity and creation of light bunches of a different form and orientation in the rooms of different function.

4. Acknowledgement
Authors express gratitude to the patent attorney of the Russian Federation V.V. Kovalenko for her assistance in drawing up the patent application [6].
The work is performed under partial support of the Russian Foundation of Basic Researches and Government of Moscow Region (project No. 17-47-500752).

References
[1] Jeffrey Y. Tsao, A fresh look at the energy economics of lighting and solid-state lighting // Journal of Physics D: Applied Physics. http://iopscience.iop.org/journal/0022-3727/page/Back%20to%20the%20future%20of%20solid%20state%20lighting.
[2] http://www.valosto.com/tiedostot/CELMA_LED(SM)054C_CELMA_position_paper_on_ADAPTORS_FINAL.pdf
[3] Gary Gordon, Interior Lighting for Designers. 5th Edition. Wiley. 2014.
[4] https://www.whichledlight.com/bulb-guide/led-bulb-beam-angle
[5] Patent for the useful model of the Russian Federation No. 100178, F21S4/00, patent holder of YUNISTAR OPTO CORPORATION (TW).
[6] Popov I V, Belyaev V V, Lekhin V , Leonisova I V. The operated light concentrator. Patent of the Russian Federation No. 116600, 20.05.2012. 10 pages.

IOP Conf. Series: Journal of Physics: Conf. Series **996** (2018) 012022 doi:10.1088/1742-6596/996/1/012022

Eulerian Circles (Venn Diagrams) as model for modern economy education on the basis of Russian professional standards

M A Sharonov[1], O V Sharonova [2], V P Sharonova[3]

[1]Moscow Region State University, 10a Radio st., 105005, Moscow, Russia
[2]Moscow State Teachers Training University, 1 Malaya Pirogovskaya st., 119435, Moscow, Russia
[3]Russian University of Cooperation, 12/30, Vera Voloshina st., 141014, Mytischi, Russia

E-mail: mik2059@yandex.ru

Abstract. The article is an attempt to create a model built using Eulerian circles (Venn diagrams) to illustrate the methodological impact of recent Federal Law 283-FZ "On the independent evaluation of qualifications" and new Federal State Educational Standards of higher education of generation 3++ on educational process in Russia. In modern economic conditions, the ability to correctly assess the role of professional standards, as a matter of fact, some set, the degree of intersection with the approximate basic educational program and the Federal State Educational Standards becomes an important factor on which in the future will depend not only the demand of graduates in the labor market, but also the possibility of passing the professional and public accreditation of the proposed program.

According to the Russian Federation law No.238-FZ "On the independent evaluation of qualifications" on the basis of approved occupational standards, created at the moment of a system of independent assessment of qualifications (hereinafter – IAQ) the employment of graduates will largely depend upon their level of skill in accordance with professional standards, or who have already been approved and are currently updating, or actively being developed. The system of independent evaluation of qualifications according to 238-FZ "on independent evaluation of qualifications" will have to earn full capacity by January 1, 2020. This fact is already taken into account in the newly developed Federal Educational Methodical Associations (hereinafter - FEMA) and already approved in the summer of this year in the Federal State Educational Standards of Higher Education of Generation 3++ (hereinafter-FSES HE 3++). Accordingly, it is now the most urgent task to adapt to these changes already developed all educational and methodological content for FSES HE 3+ depending on the degree of their intersection as multiples.

From the point of view of constructing a model using Eulerian circles at the first glance on the definition of the concepts of PS and FSES HE 3++ it becomes clear that these are different sets are not related to each other, namely [1]:

- professional standards (hereinafter PS) is characterized as a qualification necessary for the employee to make a certain type of professional activity, including perform a specific job function;

- the educational standard is characterized as a set of mandatory requirements for the formation of a certain level and (or) to the profession, specialty and the direction of training approved by the Federal Executive authority, carrying out the functions of public policy and legal regulation in the field of education.

Of course, something they must be linked methodologically. It would be easier if in the definition of Generalized Labor Functions (GLF) is also featured HE the set of concepts similar to the concepts used to define the PS, well, for example: "educational standard - a set of sequentially formed competencies required of the graduate to make a certain type of professional activity" or something like that, because it would simplify the problem of integration of the aforementioned documents, and also made it easier to use the GLF PS, as a criterion of indicators, for example, professionally-public accreditation. And here the important role is played by the General Professional Educational Program (GPEP) of FEMA and the GPEP of the University, especially since their roles are defined by the order of the Ministry of Education and Science of the Russian Federation No. 301 of 05.04.2017 [2].

In particular the paper draws attention to the fact that it is necessary to focus on GPEP. GPEP is included in the registry, and to enter this environment, you can use the front office (specialized Navigator GPEP, GPEP designer, etc.) i.e. specially designed software.

We would like to note that the development of the University's GPEP must begin with the realization that the system of independent evaluation of qualifications is, in fact, a kind of independent qualification "exam" for the educational program as a whole. By the way, this approach makes it clear that the quality of training of University specialists, their demand, on the one hand, and, on the other hand, the success of their activities in the professional sphere, the level of productivity, etc. will depend on the degree of conformity of these documents to the requirements of professional standards and how the derived requirements system IAQ [3]. To illustrate the mutual influence of these derivatives on each other can be represented as the scheme shown in figure 1.

Here becomes relevant and the necessity of taking into account the formation of Funds of Assessment Tools (hereafter FAT) to GPEP Complex Assessment Tools (hereafter CAT) was developed by experts from the IAQ profile of the Council for Professional Qualifications (hereafter CPQ), such as possible criteria for the state final attestation. The use of General Job Functions (hereafter GJF) or job functions (hereafter JF) is written in PS as indicators of achievement of these or those professional competences for the application areas of training [1; 4]. You may assume that the implementation of this approach will have a positive impact on the passage of professional and public accreditation (hereafter PPA) educational programs in the specified area of training now undertaken by the relevant CPQ.

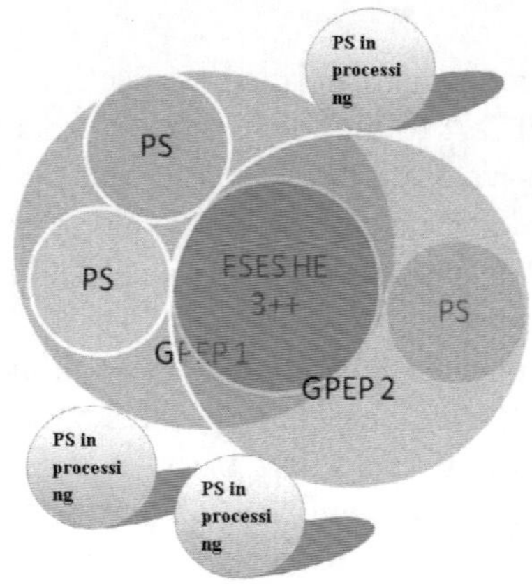

Figure 1 - Model (Euler circles) the formation of a GPEP on based PS and GEF PS is the professional standard developed by profile CPQ; FSES HE 3++ are the standards of Higher Education of Generation 3++ approved in the summer of 2017, GPEP is the approximate main educational program on a profile of preparation.

As practice shows, the role of the PS increases dramatically, if the GPEP can be formulated generalized, if there is either no corresponding updated professional standard or the direction of training combines technologically difficult compatible fields of activity. Accordingly, the task of formulating professional competencies, with the exception of their mandatory part prescribed in the GPEP fall on the shoulders of the teaching staff of the University and here simply cannot do without the PS and close interaction with the profile of the CPQ and FEMA, as shown in figure 1.

If we talk about the role of the PS, in any case, updating the developed GPEP using the quintessence of their content CAT as the accounting of practical experience and highly specialized knowledge of experts of the IAQ, for translation into the language of "competencies" adopted in high school, will be a highly relevant task.

And here we would like to share my methodological approach to solving this problem. Namely, the work function, so to say their "content" it is possible to correlate with the conceptual apparatus underlying the design of working curriculum of the discipline of the University's GPEP, and then, further and verify through the use of language "synonyms" in the description of the requirements for the employment functions PS: the "required knowledge" (wording from the professional standard) - "to know" (the wording in working curriculum of the discipline of the University's GPEP)"; "necessary skills (wording from the professional standard) - "to be able" (wording in working curriculum of the discipline of the University's GPEP)"; "labor action" (wording from the professional standard) - "possess skills" (wording in working curriculum of the discipline of the University's GPEP). Then the requirements laid down in the PS will affect the content of already working curriculum of the disciplines, and hence the process of interaction of participants in this process, their role in it and the tasks will be logically clearer and more concise. This approach will simplify the process of creating content based on the GPEP.

Here it is worth noting that for applied training areas it will be necessary to solve with the help of PS and changes in methodological approaches to the formation of programs of practices, for example, due to the need to conduct practices at a particular enterprise commissioned by a particular employer. So requirements connected, for example, with specifics of the concrete equipment or connected with frequent change (updating) of technological processes can be put forward. This may also be due to the possible changes in the introduction of the University opop new competence that allows to carry out professional activities on a specific GLF in several related PS and the existence of an order from a particular employer, etc. [1; 5]

It is this interpenetration of the requirements of the PS in the educational program will enable educational organizations carrying out preparation of experts for application areas of the economy to maintain the competitiveness of the proposed educational services, especially as in recent times training is conducted on a commercial basis.

Thus, if we draw Parallels with the proposed model of the Eulerian Circles, then the future of our graduates, their demand in the labor market, and, in fact, the demand for the relevant educational programs of the University as a whole, will depend on the correct and timely assessment of the importance of the General Professional Educational Program of Federal Educational Methodical Association a tool of the PS influence on the educational trajectory.

References

[1] Sharonov M A, Sharonova O V. On the problems of integration of professional standards and systems of higher education. Engine. 2017. 4 (112). pp. 56-57

[2] The order of the Ministry of Education and Science of the Russian Federation of 05.04.2017 № 301 on approval of the organization And implementation of educational activities on educational programs of higher education - undergraduate programs, programs of specialization, master's programs. Access method: https://rg.ru/2017/07/19/minobr-prikaz301-site-dok.html / access date on 20.11.2017.

[3] Sharonov M A, Sharonova V P Interrelation of individuality of the offered services and competitive advantages of service process. Fundamental and applied research of the cooperative sector of the economy. 2016. No. 1. C. 111-115.

[4] Reliability of service systems and equipment. Textbook // Gerasimov M K, Korosteleva V P, Sharonov M A, Sharonov V P - Moscow: Heating Engineering, 2011. - 240 p.

[5] Sharonov M A, Sharonova V P Trade business: Commerce, marketing, management. Theory and practice. Monograph / edited by doctor of Economics, prof., D I Valigursky.- M.: Publishing and Trading Corporation "Dashkov and K", 2017. – 410 p.

Institute of Physics
Dirac House, Temple Back
Bristol BS1 6BE UK

ISSN: 1742-6588
ISBN 978-1-5108-6125-1

CURRAN ASSOCIATES INC.
proceedings.com

7th International Workshop on New Computational Methods for Inverse Problems (NCMIP 2017)

Journal of Physics: Conference Series Volume 904

Paris-Saclay, France
12 May 2017

IOP
Institute of Physics